Transport and Remediation of Subsurface Contaminants

ACS SYMPOSIUM SERIES **491**

Transport and Remediation of Subsurface Contaminants

Colloidal, Interfacial, and Surfactant Phenomena

David A. Sabatini, EDITOR
University of Oklahoma

Robert C. Knox, EDITOR
University of Oklahoma

Developed from a symposium sponsored
by the Division of Colloid and Surface Chemistry
of the American Chemical Society
at the 65th Annual Colloid and Surface Science Symposium,
Norman, Oklahoma,
June 17–19, 1991

American Chemical Society, Washington, DC 1992

04789465

SEP/AE
CHEM

Library of Congress Cataloging-in-Publication Data

Transport and remediation of subsurface contaminants: colloidal, interfacial, and surfactant phenomena / David A. Sabatini, editor, Robert C. Knox, editor.

　　p.　　cm.—(ACS symposium series, ISSN 0097–6156; 491)

"Developed from a symposium sponsored by the Division of Colloid and Surface Chemistry at the 65th annual Colloid and Surface Science Symposium, Norman, Oklahoma, June 17–19, 1991."

Includes bibliographical references and indexes.

ISBN 0–8412–2223–1

1. Hazardous　　substances—Environmental　　aspects—Congresses. 2. Water, Underground—Pollution—Congresses. 3. Soil pollution—Congresses. 4. Transport theory—Congresses. 5. Colloids—Congresses. 6. Surface active agents—Congresses.

　　I. Sabatini, David A., 1957–　　. II. Knox, Robert C. III. American Chemical Society.　　Division　of　Colloid　and　Surface　Chemistry. IV. Colloid and Surface Science Symposium (65th: 1991: Norman, Okla.) V. Series.

TD427.H3T73　1992
628.5′2—dc20

92–11468
CIP

The paper used in this publication meets the minimum requirements of American National Standard for Information Sciences—Permanence of Paper for Printed Library Materials, ANSI Z39.48–1984. ∞

SD 6-16-92

Foreword

THE ACS SYMPOSIUM SERIES was founded in 1974 to provide a medium for publishing symposia quickly in book form. The format of the Series parallels that of the continuing ADVANCES IN CHEMISTRY SERIES except that, in order to save time, the papers are not typeset, but are reproduced as they are submitted by the authors in camera-ready form. Papers are reviewed under the supervision of the editors with the assistance of the Advisory Board and are selected to maintain the integrity of the symposia. Both reviews and reports of research are acceptable, because symposia may embrace both types of presentation. However, verbatim reproductions of previously published papers are not accepted.

Contents

REVIEW AND FUTURE DIRECTIONS

INDEXES

Preface

CHEMICAL RELEASES INTO THE SUBSURFACE are pervasive environmental problems. Sources of chemical releases range from abandoned hazardous waste disposal sites, such as Superfund sites, to leaking underground storage tanks at the corner gasoline station. Predicting the transport of contaminants in the subsurface and remediation of contaminated soils and groundwater has proven to be extremely challenging. Accurate prediction of subsurface contaminant transport is important in assessing the risk of public exposure to the contaminants and in evaluating various remediation scenarios. Experience has shown us that an improved understanding of subsurface contaminant transport mechanisms is needed. For example, colloidal enhanced contaminant transport may help explain instances in which contaminants have migrated much further than predicted when this process was not considered. Experience has also shown us that current remediation efforts are not satisfactory relative to the time and cost of remediation. For example, remediation is frequently inhibited by our inability to extract the contaminants from the subsurface because of the significant sorption of strongly hydrophobic chemicals, such as polychlorinated biphenyls, or because of the presence of separate phases of nonaqueous-phase liquids, such as trichloroethylene. The use of surfactants to enhance the solubility or to mobilize contaminants appears to be a promising approach that could significantly reduce the time and cost of remediation for sites contaminated with strongly hydrophobic or nonaqueous-phase liquid contaminants.

Colloidal, interfacial, and surfactant phenomena are central to reaching a more complete understanding of contaminant transport and for developing and implementing advanced subsurface remediation technologies. This book contains recent research results from leading experts in subsurface contaminant transport and remediation. An exciting aspect of this book is that it combines into one volume the results of researchers from a variety of disciplines that would typically be published in a multitude of volumes. Also, the last chapter includes the outcome of a panel and group discussion on the general topic of "Where do we go from here?" Thus, this volume combines a timely review of our current understanding of subsurface contaminant transport and remediation with an insightful discussion of critical issues that demand future consideration. This book will thus be invaluable for scientists and engineers with research, teaching, consulting, and regulatory and management responsibilities.

We thank the cochairmen of the symposium on which this book is based, John Scamehorn and Jeffrey Harwell, for their assistance in organizing the sessions and for their encouragement in producing this volume. We give special thanks to our two keynote speakers, Suresh Rao and Walter Weber, Jr., for their timely and informative presentations. We also thank the presenters at the symposium and the authors of the chapters in this book for their significant contributions, their timely submittals, and their patience. We acknowledge the members of the discussion panel (Robert Puls, John Wilson, Suresh Rao, and John Westall, the moderator) for helping to make it a most profitable exercise. We thank the reviewers, who were most cooperative and made valuable contributions to the quality of the chapters despite the short deadlines. We also thank the School of Civil Engineering and Environmental Science and the University of Oklahoma for being supportive of this activity.

Finally, we thank the ACS Books Department staff, especially Cheryl Shanks, A. Maureen Rouhi, Anne Wilson, and Paula M. Bérard, for their assistance, patience, and encouragement in preparing this book and publishing it in a timely manner.

DAVID A. SABATINI
ROBERT C. KNOX
University of Oklahoma
Norman, OK 73019

February 3, 1992

Chapter 1

Transport and Remediation of Subsurface Contaminants
Introduction

Robert C. Knox and David A. Sabatini

School of Civil Engineering and Environmental Science, University of Oklahoma, Norman, OK 73019

Contamination of subsurface soils and ground water formations is a pervasive environmental problem that has proven to be extremely difficult to remediate. Cleanup of contaminated subsurface environments is complicated both by the physical nature of the formation and the behavior of contaminants introduced to the formation. Soils and many aquifers are porous media comprised of solid material and pore spaces which are occupied by fluids and/or gases. In addition to a complex physical structure, subsurface formations are hidden from view and highly inaccessible. With accessibility limited to localized points (boreholes) or complete disruption (excavation), remediation of the subsurface environment represents a formidable challenge.

The myriad of materials that have been introduced to the subsurface environment have resulted in excessive levels of heavy metals, organic and inorganic chemicals and bacteriological agents. The contaminant materials can exist in the subsurface in different phases; attached to the soil solids, dissolved in the ground water, or occupying the pore spaces as a separate gaseous or liquid phase. Properties of the phase(s) in which contaminant materials exist in the subsurface influence the potential mobility of the contaminants. This chapter will provide introductory comments relative to contaminant transport and fate and subsurface remediation and background information relative to the book and the ACS Symposium Session which served as the catalyst.

Transport and Fate Processes

The two basic elements affecting the transport and fate of contaminants in the subsurface are properties of the subsurface environment itself and physicochemical and biological properties of the contaminant. Nonreactive (conservative) chemicals tend to move unimpeded through the subsurface and subjected only to hydrodynamic processes. Conservative chemicals are not affected by abiotic or

0097–6156/92/0491–0001$06.00/0

biotic processes which can occur in the subsurface. Conversely, nonconservative contaminants can be affected during ground water transport if subsurface conditions are conducive to potential reactions. Thus, for interactions between the subsurface environment and the contaminant to occur, it is necessary that both the contaminant property and the subsurface environment be conducive to these interactions. Relatively simple examples of hydrodynamic, abiotic and biotic processes are presented in this section; facilitated transport and multiphase flow are described in subsequent sections.

Table 1 gives an expanded list of subsurface processes and corresponding subsurface and contaminant properties influencing these processes. The general categories of processes affecting subsurface fate and transport are hydrodynamic processes, abiotic (nonbiological) processes and biotic processes. Hydrodynamic processes affect contaminant transport by impacting the flow of ground water (in terms of quantity of flow and flow paths followed) in the subsurface. Examples of hydrodynamic processes are advection, dispersion, preferential flow, etc. The hydrodynamic process of dispersion (spreading of the contaminant about the mean ground water velocity) is attributed to the distribution of flow paths for the subsurface system and the diffusion of the contaminant. Dispersivity is defined as a soil parameter which describes the spreading due to the distribution of flow paths (when Fickian dispersion is assumed). However, for low ground water flow velocities, molecular diffusion may dominate the contaminant transport. Thus, the dispersivity and ground water velocity of the subsurface system, and the molecular diffusion coefficient of the contaminant, are properties that control the dispersion process.

Abiotic processes affect contaminant transport by causing interactions between the contaminant and the stationary subsurface material (e.g., adsorption, ion exchange) or by affecting the form of the contaminant (e.g,, hydrolysis, redox reaction). The abiotic process of adsorption (accumulation of the contaminant at the surface of a solid interface - typically the stationary subsurface material) will slow down the movement of the contaminant as it accumulates on the subsurface medium. For neutral organic contaminants and subsurface materials with organic carbon content present, the adsorption is commonly of hydrophobic (water disliking) chemicals into organic carbon content. The more hydrophobic a chemical, the greater the water disliking characteristic of the chemical. Thus, as the solubility of a chemical decreases, the potential for the chemical to adsorb at an interface is expected to increase. Also, as the organic carbon content of the subsurface material increases, the total capacity of the soil to adsorb the contaminant increases. This will result in additional ground water passing through the subsurface material before the adsorptive capacity is exceeded and the contaminant appears downgradient. Thus, the organic carbon content of the subsurface system and the solubility (hydrophobicity) of the contaminant are two properties that affect the adsorption process. As the organic carbon content decreases and/or if metals are of concern, the aqueous chemistry (pH, ionic strength, etc.) and nature of the mineral surfaces become increasingly important.

Biotic processes can affect contaminant transport by metabolizing or mineralizing the contaminant (e.g., organic contaminants) or possibly by utilizing the contaminant in the metabolic process (e.g., nutrients, nitrate under denitrifying conditions). As an example, the biotic process of aerobic biodegradation may convert an organic contaminant to another form (metabolites) or to harmless end products (e.g., CO_2 and H_2O). For aerobic biodegradation (metabolism) to function, aerobic microorganisms must be present in the subsurface system. The microorganisms require free oxygen as an electron acceptor (free oxygen is present only at high values of pE), nutrients (such as nitrogen and phosphorous), certain trace elements and an acceptable environment (pH, temperature, etc.). The contaminant must be readily metabolized (e.g., highly halogenated organic contaminants are typically refractory under aerobic conditions - difficult for microorganisms to metabolize) and present at sufficient concentrations to make the metabolism energetically favorable. The biochemical oxygen demand (BOD) is a test that quantifies the biochemical oxygen equivalent of the organics present in a contaminant and thus indicates if the contaminant is susceptible to aerobic biodegradation. Thus, the aerobic biodegradation process is affected by properties of the subsurface environment (presence of microorganisms, nutrients, free oxygen, etc.) and properties of the contaminant (concentration, BOD). The process of anaerobic biodegradation is also affected by both subsurface environment and contaminant properties.

Facilitated Transport

The term "facilitated transport" is used to describe those phenomena whereby a contaminant is transported at rates beyond those expected based on consideration of Darcian flow and equilibrium sorptive reactions. The term can be considered somewhat of a misnomer in that it implies that the contaminant is transported faster than it should be. On the contrary, the unexpected rate of movement of the contaminant is the result of an incomplete consideration of the possible transport and fate processes.

Facilitated transport processes are broadly grouped into two categories; cosolvent effects and colloidal processes. According to Huling (1989), cosolvent effects are attributable to the presence of a miscible organic solvent in addition to the aqueous solvent (ground water) which effectively reduces the polarity of the mixture and thus the solvophobicity of the contaminants. Hydrophobic organics will partition more strongly into the solvent mixture and thus exhibit increased mobility (due to less sorption and retardation).

Traditional consideration of solute transport processes is based on a mass balance of the solute between the mobile aqueous phase and the immobile solid phase. Under certain conditions, small solid phase particles and macromolecules, which exist in some subsurface environments, are transported in the aqueous phase. These mobile sorbents (referred to as colloids based on their size) can sorb contaminants, thus causing more significant migration of strongly sorbing chemicals (Huling, 1989).

Multiphase Behavior

Increasingly, organic liquids with limited aqueous solubilities are being released to subsurface environments and threatening ground water resources. Organic

Table 1. Subsurface Processes and Corresponding Subsurface and Contaminant Properties and Interactions Affecting the Fate and Transport of Contaminants

Process	Subsurface Property	Contaminant Property	Interactions
Hydrodynamic Solute Transport			
Advection	Ground water gradient, hydraulic conductivity, porosity	Independent of contaminant	
Dispersion	Dispersivity, pore water velocity	Diffusion coefficient	Dispersion coefficient
Preferential Flow	Pore size distribution, fractures, macropores		
Abiotic Solute Transport			
Adsorption	Organic content, clay content, specific surface area	Solubility, octanol-water partition coefficient	
Volatilization	Degree of saturation	Vapor pressure, Henry's constant	
Ion Exchange	Cation exchange capacity, ionic strength, background ions	Valency, dipole moment	

Hydrolysis	pH, competing reactions	Hydrolysis half life	
Precipitation/ Dissolution	pH, other metals	Solubility versus pH, speciation reactions	
Cosolvation	Types and fraction of cosolvents present	Solubility, octanol-water partition coefficient	
Redox	pE, pH	pK_a	
Colloid Transport	pH, ionic strength, flow rate, mobile particle size, aquifer and particle surface chemistry	Sorption, reactivity, speciation, solubility	Colloid stability
<u>Biotic</u>			
Metabolisim/ Cometabolism	Microorganisms, nutrients, pH, pE (electron acceptors), trace elements	BOD, COD, degree of halogenation, etc.	
Multiphase Flow	Intrinsic permeability, saturation, porosity	Solubility, volatility, density, viscosity	Relative permeability, residual saturation, wettability, interfacial tension (surface tension), capillary pressure

liquids with limited aqueous solubilities are often referred to as immiscible with respect to the aqueous phase. The terms miscible and immiscible actually refer to the absence or presence of capillary pressures between two fluids. Immiscible fluids would be those that have significant capillary pressures resulting from interfacial surface tension effects.

Because of the widespread occurrence of fluids immiscible with water general terms and acronyms have been promulgated. The general term for the immiscible fluids is Non-Aqueous Phase Liquids (NAPLs). Generally, NAPLs are divided into two classes based on density: those that are lighter than water (LNAPLs); and those with a density greater than water (DNAPLs). Most natural and refined hydrocarbons are LNAPLs, while most chlorinated organics and solvents are DNAPLs.

In considering the transport of NAPLs in the subsurface, density and viscosity are of prime interest (Corapcioglu and Hossain, 1986). Density of an immiscible organic fluid is the parameter which delineates LNAPLs from DNAPLS, i.e. the "floaters" from the "sinkers". Viscosity is a measure of a fluid's resistance to flow. A less viscous fluid will more readily penetrate a porous media. Fluids with higher densities and lower viscosities than water will be more mobile in the subsurface than water (Huling and Weaver, 1991).

NAPL migration in the subsurface is affected by the characteristics of the release scenario (volume of release, area of infiltration, duration), properties of both the NAPL and subsurface media, and subsurface flow conditions (Feenstra and Cherry, 1987). For LNAPLs released to the subsurface, the product will migrate downward only if sufficient amount is available to meet residual saturation needs of the soil material. The product continues downward until the capillary fringe is reached, at which point the product will tend to mound. The increased head will depress the water table. The product can also move laterally due to capillary forces in the vadose zone (Tyler et al. 1987).

For DNAPLs released to the subsurface, the product will move downward through the subsurface in a non-uniform pattern. Viscous fingering occurs due to the unstable flow of DNAPLs through water. At the capillary fringe, the DNAPL will mound slightly to build enough head to displace the water held by capillary forces to the soil medium. The DNAPL will continue downward through the saturated zone until residual saturation is reached or an impermeable soil zone is encountered. After ponding on the impermeable layer, the DNAPL will spread out laterally due to gravitational forces (head differences) and due to viscous drag exerted by the flowing ground water (Parker et al. 1986).

It should be noted that LNAPLs and DNAPLs are not totally immiscible. Some of the product goes into solution in the saturated zone. In subsurface flow of immiscible fluids, the possibility also exists that a gaseous phase could occupy part of the available pore spaces. Therefore, in the unsaturated zone, the transport of a contaminant could include three phases: solutes in the aqueous phase; vapors in the air phase; and unaltered constituents in the immiscible phase (Corapcioglu and Hossain, 1986).

In the saturated zone below the water table, immiscible contaminant transport is usually assumed to involve two phases; the aqueous phase and the immiscible phase. In most work, these two phases are assumed to be distinct and each behaves as a separate continuum (Reible et al. 1986). The complexity of the mathematical statement increases significantly when the number of phases increases.

Subsurface Remediation

Remediation of subsurface contamination has proven to be problematic. Early efforts at ground water cleanup were characterized as being costly, time-consuming and ineffective (Knox et al 1986). Recent years have seen dramatic increases in the number of technologies being promoted for subsurface remediation. Many of the new technologies represent simple innovations of existing procedures. A broad categorization of remediation technologies could include (USEPA, 1990):

1. Pump and Treat - extraction of contaminated ground water with subsequent treatment at the surface and disposal or reinjection.

2. Soil Vacuum Extraction - enhanced volatilization of compounds by applying a vacuum to the subsurface.

3. Soil Flushing/ Washing - use of extractant solvents to remove contaminants from soils.

4. Containment - emplacement of physical, chemical, or hydraulic barriers to isolate contaminated areas.

5. Bioremediation - enhanced biodegradation of contaminants by stimulating indigenous subsurface microbial populations.

Most remediation schemes will involve a combination of technologies. However, for any technology to be effective, a thorough understanding of the processes governing the transport and ultimate fate of the target contaminants needs to be developed. With few exceptions, the above technologies are highly influenced by physicochemical and biological processes active in the subsurface environment.

Book Topic

The key points to be extracted from the discussion above are (1) the subsurface transport and fate of contaminants can be influenced by numerous processes; (2) successful remediation of subsurface contamination will require manipulation of the subsurface processes; and (3) most of the subsurface processes are influenced by physicochemical processes. Given the magnitude of subsurface contamination problems and the rather poor success rate for current remediation technologies, it is obvious that an improved understanding of the processes governing subsurface transport and fate of contaminants and limiting ground water and soil remediation is needed.

Interest in promoting the session on "Colloidal, Interfacial and Surfactant Phenomena in Subsurface Contaminant Transport and Remediation" is an outgrowth of research interests and interdisciplinary efforts of faculty from the schools of Chemistry and Chemical Engineering and Materials Science (CEMS), the Institute for Applied Surfactant Research (IASR), the school of Civil Engineering and Environmental Science (CEES), and the Environmental and Ground Water Institute (EGWI), all at the University of Oklahoma. Faculty and researchers from Chemistry, CEMS, and IASR have focused their research efforts on using surfactant technology in a variety of applications. Members of the CEES faculty and personnel from EGWI have studied subsurface transport and fate processes and aquifer restoration measures. An interdisciplinary research effort focusing on using surfactant technology for subsurface remediation identified the need for information dissemination related to the session topic.

The session topic was developed during the early planning stages for the 65th Colloid and Surface Science Symposium. A nationwide call for abstracts resulted in a total of 34 papers being accepted for presentation at the session. Table 2 lists the presentation titles with authors and affiliations. Each presentation was considered eligible for inclusion in this manuscript. Actual participation in the book was left up to the individual authors. The book includes a total of 16 session presentations and 1 invited contribution. The chapters are organized into topical groupings addressing colloids, inorganics, surfactants, and organics. The final chapter provides an overview of the book (summary of the chapters) and a summary of the panel discussion which addressed the question of where do we go from here.

Table 2.   Paper and Poster Presentations on Colloid and Interfacial Aspects of Groundwater and Soil Cleanup

Papers

1.   "Sorption and Transport of Organic Contaminants in Soils and Aquifers" by P.S.C. Rao, University of Florida, Gainesville, FL.

2.   "Distributed Sorption Reactivity: Implications for the Behavior of Organic Contaminants in Subsurface Systems" presented by L. E. Katz for Dr. Walter J. Weber, University of Michigan, Ann Arbor, MI.

3.   "Electrostatic Repulsive Effects on the Mobility of Inorganic Colloids in Subsurface Systems" by Dr. Robert W. Puls, USEPA/RSKERL, Ada, OK.

4.   "Colloid Transport in Porous Media with Repulsive Energy Barriers" by Dr. T. M. Olson, University of California, Irvine, CA.

5.   "Permeabilities of Small-pore Oxide Membranes" by Dr. M. J. Gieselmann, University of Wisconsin, Madison, WI.

Table 2. Continued

6. "Capture of Colloids in Porous Media: Theory, Numerical Solution and Implications to the Transport of Colloidal Contaminants in Ground Waters" by Dr. E. Elimelech, University of California, Los Angeles, CA.

7. "Experiments on Transport of Colloids During Gas and Aqueous Fluid Flow in Porous Media" by J. Wan, New Mexico Tech., Socorro, NM.

8. "Colloids and Immiscible Organics: How Colloids and the Physical Behavior of an Immiscible Organic Phase Leads to Greater Capillary Trapping of the Organic Phase" by R. E. Mace, New Mexico Tech, Socorro, NM.

9. "The Relationship Between Interfacial Properties and Two-Phase Flow of Organic Contaminants in Groundwater" by Dr. K. F. Hayes, University of Michigan, Ann Arbor, MI.

10. "Wetting and Residual Non-Aqueous Phase Liquid (NAPL) Saturation in Surface-Altered Uniform Glass Beads" by M. Wei, New Mexico Tech, Socorro, NM.

11. "Interfacial Mass Transfer of Actinides in NAPL-Groundwater System" by M. P. Gardiner, Harwell Laboratory, Didcot, UK.

12. "Exposure Assessment Modeling for Hydrocarbon Spills in the Subsurface: Sensitivity Analyses" by Dr. J. W. Weaver, USEPA/RSKERL, Ada, OK.

13. "Ordered Humic Aggregates as Pseudo-phases in Hydrologic Systems" by Dr. R. L. Wershaw, USGS, Arvada, CO.

14. "Landfill Leachate Effects on Transport of Organics in Aquifer Materials" by Dr. F. Pfeffer, USEPA/RSKERL, Ada, OK.

15. "Production and Partial Characterization of Selenium Colloids Produced by Bacteria" by Dr. L. L. Barton, University of New Mexico, Albuquerque, NM.

16. "Effect of Micellar Solubilization on Biodegradation Rates of Hydrocarbons" by S. J. Bury, Rice University, Houston, TX.

17. "Fundamental Considerations in the Selection of Surfactants for In Situ Applications" by Dr. J. H. Harwell, University of Oklahoma, Norman, OK.

18. "Mineralization of Phenanthrene in Soil-Water Systems with Nonionic Surfactants" by Dr. R. G. Luthy, Carnegie Mellon University, Pittsburgh, PA.

Continued on next page

Table 2. Continued

19. "Potential Use of Cationic Surfactants in Aquifer Remediation" by Dr. D. R. Burris, HQAFESC/RDVC, Tyndall AFB, FL.

20. "Modification of Mineral Surfaces with Cationic Surfactants " by Dr. J. C. Westall, Oregon State University, Corvallis, OR.

21. "Sorption of Hydrophobic and Amphiphilic (Nonionic Surfactant) Organics with Subsurface Materials" by Dr. D. A. Sabatini, University of Oklahoma, Norman, OK.

22. "Surfactant-Enhanced Solubilization of Tetrachloroethylene and Degradation Products in Pump and Treat Remediation " by Dr. C. C. West, USEPA/RSKERL, Ada, OK.

23. "A Comparison of Field Tests of Surfactant Flooding: Examples of Mobility Control of DNAPL" by Dr. J. C. Fountain, State University of New York, Buffalo, NY.

24. Colloid Remediation in Groundwater by Polyelectrolyte Capture" by R. Jain, University of New Mexico, Albuquerque, NM.

25. "The Application of Heap Leaching Technology to the Treatment of Metals Contaminated Soil" by Dr. A. T. Hanson, New Mexico State University, Las Cruces, NM.

Posters

1. "Swelling Properties of Soil Organic Matter and Their Relationship to Sorption of Organic Molecules from Mixtures and Solutions" by Dr. W. G. Lyon, USEPA/RSKERL, Ada, OK.

2. "Utilization of Surfactants for Aquifer Remediation: Effects of Groundwater Temperature and Chemistry on Surfactant Performance" by M. S. Diallo, University of Michigan, Ann Arbor, MI.

3. "Application of Gradient Elution Techniques for Assessment of Organic Solute Mobility in Solvent Mixtures" by A. L. Wood, USEPA/RSKERL, Ada, OK.

Literature Cited

1.   Corapcioglu, M.Y. and Hossain, M.A. (1986). "Migration of Chlorinated Hydrocarbons in Groundwater", *Hydrocarbons in Ground Water: Prevention, Detection and Restoration*, National Water Well Association, pp. 33-52.

2.   Feenstra, S. and Cherry, J.A. (1987). "Dense Organic Solvents in Ground Water: An Introduction", Progress Report 0863985, Institute for Ground Water Research, University of Waterloo, Canada.

3.   Huling, S. (1989). "Facilitated Transport", EPA/540/4-89/003, U.S. Environmental Protection Agency, Ada, Oklahoma.

4.   Keely, J.F. (1989). "Performance Evaluations of Pump-and-Treat Remediations", EPA/540/4-89/005, U.S. Environmental Protection Agency, Ada, OK.

5.   Huling, S.G. and Weaver, J.W. (1991). "Dense Nonaqueous Phase Liquids", *Ground Water Issue,* EPA/540/4-91-002, U.S. Environmental Protection Agency, Ada, OK.

6.   Knox, R.C., Canter, L.W., Kincannon, D.F., Stover, E.L. and Ward, C.H. (1985). "State-of-the Art of Aquifer Restoration", EPA/600/S2-84/182, U.S. Environmental Protection Agency, Ada, OK.

7.   Parker, J.C., Lenhard, R.J. and Kuppusamy, T. (1986). "Modeling Multiphase Contaminant Transport in Ground Water and Vadose Zones", *Hydrocarbons in Ground Water: Prevention, Detection and Restoration, National Water Well Association*, pp. 189-200.

8.   Reible, D.D., et al. (1986). "Development and Experimental Verification of a Model for Transport of Concentrated Organics in the Unsaturated Zone", *Hydrocarbons in Ground Water: Prevention, Detection and Restoration*, National Water Well Association, 1986. pp. 107-126.

9.   Tyler, S.W., et al. (1987). "Processes Affecting Subsurface Transport of Leaking Underground Tank Fluids", EPA/600/6-87/005, U.S. Environmental Protection Agency, Las Vegas, NV.

10.   U.S. Environmental Protection Agency (1990). "Subsurface Contamination Reference Guide", EPA/540/2-90/011, Office of Emergency and Remedial Response, Washington, D.C.

RECEIVED January 28, 1992

COLLOIDS

Chapter 2

Colloid Deposition in Porous Media and an Evaluation of Bed-Media Cleaning Techniques

T. M. Olson and G. M. Litton

Department of Civil Engineering, University of California, Irvine, CA 92717

Deposition rates of carboxyl latex spheres (0.245 μm) in packed beds were examined in the presence of repulsive electrostatic interaction forces. Quartz grains and glass beads (275 μm), were selected for comparison as bed media on the basis of their similar zeta potentials. Observed attachment efficiencies (α) were compared with values predicted from models based on DLVO theory at varying pH and ionic strength. Attachment efficiencies in quartz and glass media were similar only once aggressive cleaning measures for both surfaces were employed. Observed values of α were also much closer to theoretical predictions with this treatment than without it, although they are still generally under-estimated. Values of α as small as 3.4×10^{-4} have been measured at low ionic strengths by using columns up to 4 meters in length. The results suggest that greater colloid mobility can be attributed to electrostatic repulsion than has been previously demonstrated.

Attempts to predict colloid attachment efficiencies in the presence of repulsive electrostatic interaction energy barriers have drastically under-estimated deposition rates (1-6). While DLVO theory does not appear to adequately describe surface interactions at close separation distances, it is not yet clear whether other artifacts contribute significantly to the discrepancy. Filtration experiment results among different investigators using the same materials often differ significantly, and irreproducibility between experiments is frequently a problem. Factors, such as physical and chemical collector surface heterogeneity, have been invoked as contributing reasons for these observations. Characterization information of the collector surface is only generally available in terms of bulk parameters, however, such as streaming potential measurements.

The objective of this study was to compare the deposition rates of submicron latex spheres on two collector materials with very similar size and surface potentials, but different phase properties and surface roughness. The two selected bed media were quartz grains and soda-lime glass beads. Theoretical predictions of attachment efficiencies based on DLVO theory are indistinguishable for quartz and glass at the same pH and ionic strength. Based on the

0097–6156/92/0491–0014$06.00/0

relative rates of colloid deposition in these systems, an indication of the importance of factors other than van der Waals attraction and electrostatic repulsion was sought. The sensitivity of results to the cleaning techniques used for the bed media was also examined.

Experimental Procedures

Materials and Cleaning Techniques. Bed media in the filtration experiments consisted of uniformly-sized ultra-pure quartz grains (Feldspar Corp., now Unimin) or soda-lime glass beads (Cataphote, Inc.). Prior to cleaning, the raw quartz material required further size fractionation. Wet sedimentation and flotation techniques were used to remove quartz grains that were larger and finer, respectively, than the desired range. This method was adopted over mechanical sieving techniques, since sieving was found to introduce surface contaminants that are not easily removed. Subsequent cleaning procedures for the quartz consisted of the following steps: 1) soaking in concentrated HCl for 24 h, 2) flushing in a rinsing column for 2 h, 3) drying at 100° C, 4) calcining at 810° C for 8 h, 5) partial cooling in the furnace for 8 h and then to room temperature in a vacuum dessicator, 6) boiling in 4 N HNO_3 for 30 min, and 7) batch-washing until the pH was approximately 5.6. The final boiling step served to deaerate the material. Experiments indicate that the final use of nitric acid is probably not necessary, however, it was used as a precautionary measure to remove any organics that might have condensed on the surface during the cooling step. Once clean, the quartz was stored wet for only short periods before it was used.

Heat treatment at the high temperatures used above, however, could not be used for the glass beads. Instead, the heat treatment was replaced with soaking in concentrated HCl, then in concentrated chromic acid for 24 h, followed by soaking in concentrated HCl, rinsing, and storing wet. Prior to use, the beads were also boiled in 4 N HNO_3 for 30 min.

The colloids were surfactant-free carboxyl latex spheres that had been ion-exchanged by the manufacturer (Interfacial Dynamics Corp.). Before use, the latex spheres were batch-washed with deionized water and centrifuged 8-10 times until the specific conductivity was less than 2 μmhos cm^{-1}.

Solutions were prepared with analytical grade reagents and Millipore Milli–Q water (18 MΩ cm), and were filtered through a 0.050 μm membrane filter. All glassware and column sections were pre-washed with chromic acid, rinsed with deionized water, concentrated HCl, and finally deionized water again.

Material Characterization Tests. Surface charge characteristics and the zero point of charge (pH_{zpc}) of the latex spheres were determined by conductometric and potentiometric titrations, respectively. Zeta potential values and the isoelectric point (pH_{iep}) were calculated from electrophoretic mobility measurements made with a Rank Bros. Mark IV apparatus. Zeta potentials for a range of ionic strengths (0.001 to 0.1 M) are shown in Figure 1. Particle size and dispersity were checked by photon correlation spectroscopy (Coulter N4MD apparatus) and these results agreed closely with the specifications of the manufacturer. The characteristics of these colloids are summarized in Table I.

The surface zeta potentials of the quartz and glass were estimated from electrophoretic mobility measurements of crushed quartz and glass. The quartz and glass media were ground in an agate mortar and particles with approximate diameters of 1.27 and 1.50 μm, respectively, were obtained by sedimentation techniques. Both materials were cleaned by the methods described above for the bulk material. Our experimental electrokinetic zeta potential measurements are compared with reported streaming potential data as a function of pH in Figure 2. Both methods indicate that glass and quartz surfaces have very similar

Table I. Colloid and Collector Properties

	Latex Spheres	Quartz Grains	Glass Beads
Surface groups	carboxyl		
Mean diameter (μm)	0.245($\pm$ 22)	250-300	250-300
Density (g cm^{-3})	1.055	2.68	2.45
Spec. surface area (m^2 g^{-1})	23.2	N.D.[a]	N.D.
Spec. surface charge (μC cm^{-2})	9.5($\pm$ 0.5)	N.D.	N.D.
Area per site[b] (nm^2)	1.7	N.D.	N.D.
pH$_{zpc}$	3.8($\pm$ 0.1)	N.D.	N.D.
pH$_{iep}$[b]	3.7	3.0	3.0

[a]N.D. = Not determined.
[b]Values determined at an ionic strength of 0.001 M for glass
and 0.001 - 0.1 for latex and quartz.

surface potentials. The agreement between the streaming potential data and electrokinetic data for the crushed quartz suggests that the crushing step did not detectably affect its surface charge characteristics.

Filtration Experiments. Glass-walled columns (1.59 cm I.D.) were packed to achieve near constant porosities; 0.390 ($\pm$ 0.010) and 0.480 ($\pm$ 0.013) for glass bead and quartz grain column media, respectively. Various column lengths between 0.025 to 4 m were employed to obtain acceptable resolution of particle collection rates. Only final values of relative concentrations (C/C$_0$) between 0.12 and 0.97 were considered acceptable. In each experiment, effluent concentrations were measured for periods equivalent to at least 8 pore volumes.

Prior to each experiment, the media was flushed at the same flow rate as the experiment with a background electrolyte solution that had been prepared fresh. The pH of the effluent was monitored in a flow-through cell until its value was within 0.01 units of the influent pH. Equilibration in the quartz columns was rapidly attained. Significantly longer flushing times were required for the glass bead media, due to its higher solubility and the presence of cations associated with soda and lime.

Colloidal suspensions were pumped downward through the bed media with a peristaltic pump at 3.50 ml min^{-1} (the corresponding approach velocity is 0.03 cm s^{-1}). The influent particle concentration was 0.25 mg L^{-1}. Particle concentrations in the column effluent were determined from UV-light scattering correlations. Using a double-beam spectrophotometer (Hitachi, Model U2000), light absorbance changes at 234 nm were measured in a flow-through cell with an optical path of 10 cm. The hydraulic retention time in the cell was 150 s.

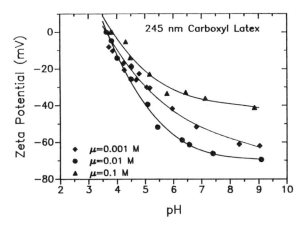

Figure 1. Microelectrophoresis determinations of surface zeta potentials vs pH for carboxyl latex spheres at varying ionic strengths.

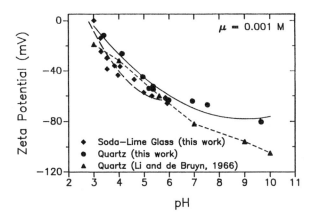

Figure 2. Comparison of electrokinetic zeta potential measurements for crushed quartz and soda-lime glass with streaming potential measurements from the literature.

Below pH 5.6 and above pH 8.0, the pH was adjusted with HCl or NaOH, respectively. In the neutral pH regime, suspensions were buffered by appropriate additions of $NaHCO_3$ and by equilibration with atmospheric CO_2. During filtration experiments, the pH of the effluent was continuously monitored in a flow-through pH cell. For experiments at high pH, the atmosphere in the reservoir containing the suspension was kept CO_2-free with the use of an Ascarite tube.

Data Analysis and Modeling Approach

Experimental Attachment Efficiencies. In the absence of repulsive interaction forces, classical descriptions of colloid deposition rates are formulated in terms of single collector efficiencies, η_0, that depend on the basic collection mechanisms of particle diffusion, interception, and sedimentation (8). When repulsive energy barriers exist, the deposition rates decrease by an attachment efficiency factor, α, defined as:

$$\alpha = \frac{\text{particle attachment rate with repulsive barriers}}{\text{particle attachment rate without repulsive barriers}}. \quad (1)$$

The overall dimensionless collection efficiency, η_t, that can be experimentally determined, therefore, is equal to the product, $\eta_0\alpha$. Using filtration experiment breakthrough curves during initial deposition conditions and a mass balance analysis of particles, "experimental" values of α may be calculated as:

$$\alpha_{exp} = -\ln(C/C_0)\left[\frac{4a_c}{3(1-f)L\eta_0}\right], \quad (2)$$

where C/C_0 is the plateau value of the relative effluent concentration, f is the bed porosity, L is the bed depth, and a_c is the collector radius. A theoretical expression for η_0 is required for this calculation and the formulation derived by Levich (9) was used:

$$\eta_0 = (18)^{\frac{1}{3}}A_s^{\frac{1}{3}}\left[\frac{D}{Ua_c}\right]^{\frac{2}{3}}, \quad (3)$$

where, A_s is a porosity factor derived by Happel (10), and U is the approach velocity at infinity.

Typical breakthrough curves for filtration experiments in quartz bed media are shown in Figure 3. The plateau value of C/C_0 remained constant in well-cleaned media, but tended to increase at a very gradual rate with less-efficiently cleaned media.

Modeling Approach. The transport equation for deposition rates of colloidal particles on spherical collectors is:

$$\frac{\partial C}{\partial t} + \mathbf{v}\cdot\nabla C = \nabla\cdot\left[D\nabla C + \frac{DC}{kT}\nabla\Phi_t\right], \quad (4)$$

where D is the particle diffusion coefficient in bulk solution, k is Boltzmann's constant, T is the temperature, t is time, and Φ_t is the total potential interaction energy. An approximate solution to equation 4 has been developed by Spielman and Friedlander (11). Their approximation for η_t:

$$\eta_t = \eta_0 \left[\frac{\beta}{1 + \beta} \right] S(\beta) , \qquad (5)$$

is valid when repulsive forces exist. The term $S(\beta)$ is a complex function for which values have been tabulated (11). The non-dimensional parameter β reflects the relative importance of diffusion-controlled particle collection versus "reaction-controlled" collection and is given by:

$$\beta = \frac{(2)^{\frac{1}{3}}}{3} \Gamma \left[\frac{1}{3} \right] A_s^{-\frac{1}{3}} \left[\frac{D}{U a_c} \right]^{\frac{1}{3}} \left[\frac{k' a_c}{D} \right] , \qquad (6)$$

where k' is an effective first-order surface rate constant that has been modified by Dahneke (12) to consider hydrodynamic interactions at close separation distances. Dahneke's solution for k' is a function of the total potential interaction energy and the separation distance, h, as follows (in approximate form):

$$k' = \frac{D}{\int_0^\infty \left\{ \left[1 + \frac{a_p}{h} \right] \exp(\Phi_t/kT) - 1 \right\} dh} . \qquad (7)$$

Based on equation 5, the theoretical attachment efficiency can be calculated as:

$$\alpha = \left[\frac{\beta}{1 + \beta} \right] S(\beta) . \qquad (8)$$

The function $S(\beta)$ is a much less-sensitive function of β than the term, $\beta/(1+\beta)$. In addition, when the maximum potential interaction energy is much greater than 25 kT, β, and hence α, go to zero rapidly. It is this sharp dependence on interaction energies that has not been observed in filtration experiments. Even with the use of well-cleaned bed media, we will show that significant differences between theoretical and observed α values still remain.

Results and Discussion

For the colloids and collector materials used in this study, unfavorable deposition conditions are expected when the pH is greater than the isoelectric point of the latex spheres. In this range both the collector and colloid surfaces are negatively charged. Using the zeta potential measurements in Figures 1 and 2, and classical solutions for the electrostatic double layer repulsive force between two constant potential surfaces (13), the interaction energies, Φ_r, were estimated. Interaction energies due to London-van der Waals attraction, Φ_a, were calculated using the expressions developed by Gregory (14). Example profiles of the total potential interaction energies, $\Phi_t = \Phi_r + \Phi_a$, are illustrated in Figure 4 at varying pH.

Only small differences are predicted between the interaction energies at

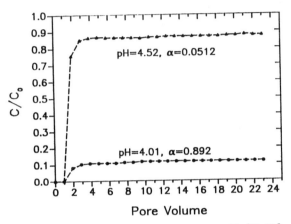

Figure 3. Typical breakthrough curves at varying pH, [NaCl] = 0.01 M, L = 10 cm, U = 0.03 cm s^{-1}.

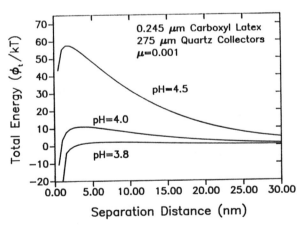

Figure 4. Potential interaction energies vs separation distance for carboxyl latex spheres and a quartz grain collector at varying pH.

ionic strengths of 0.001 and 0.01 M. This prediction is due to the anomalous surface potential behavior of latex particles at low ionic strengths (see Figure 1). Maximum surface potentials for latex spheres have consistently been reported as a function of ionic strength (15-17). Qualitative models describing this behavior have been proposed. One model, known as the "hairy layer" model, attributes the decreased mobility in weak electrolyte to the presence of long polymer molecules, or "hairs", that allow the charged end groups to extend out from the surface (15). The compression or extension of these polymer chains changes the position of the shear plane, and hence the electrophoretic mobility of the sphere. Other proposed mechanisms involve the non-specific association of co-anions (16,17). The potential effects that each of these mechanisms would have on particle deposition rates are not necessarily equivalent, and in fact, filtration experiments may help to elucidate the mechanism.

Deposition Rates in Well-Cleaned Media. The rates of colloid attachment in the quartz, when prepared by the procedures outlined above, are shown in Figure 5. At ionic strengths of 0.01 M or greater, the observed and theoretically-predicted onset of unfavorable deposition exactly coincide. When the ionic strength was reduced to 0.001 M, however, the data suggest the presence of repulsive interaction forces at lower pH values than predicted by theory. Of those mechanisms proposed to date, the "hairy layer" model can most easily explain this observation. In a microelectrophoresis cell, the dangling charged, polymer molecules at the surface of a latex colloid would be extended in sufficiently weak electrolyte. If the same colloid were to approach another negatively charged surface, the polymer chains would become compressed and the latex surface zeta potential would appear to increase. Their compression, however, may require overcoming negative entropic forces, and therefore, additional factors such as steric repulsion may be significant.

Experiments conducted under favorable deposition conditions in quartz media suggest that the observed value of η_0 is approximately 20% greater than the theoretical value. Only very minor shifts in the data of Figure 5 would result, therefore, if the true value of η_0 had been used. Similar close agreement with theoretical predictions for η_0 in beds of irregularly shaped media, such as sand grains, is widely reported (18).

Colloid deposition rates in "well-cleaned" glass and quartz bed media (see Figure 6) are relatively similar, although in general the deposition rates in glass bead media are slightly greater. Since the surface zeta potentials for glass are slightly more negative than quartz, slower deposition might have been expected. These differences may be indicative of residual impurities on the glass surface. Since the amounts of impurities in the bulk glass material are greater than those in the quartz, and the glass is more soluble, it may not be possible to completely "clean" the glass bead surface.

While substantial differences still remain between the predicted and experimental values of α, much greater α values were measured for most of the other bed media cleaning procedures that were used. In addition, similar deposition rates in quartz and glass media were only observed after scrupulous cleaning of these materials. With the elimination of this experimental artifact, the differences between experimental and theoretical observed attachment efficiencies, are more likely now to be corrected by modeling approaches.

Comparison of Cleaning Techniques. Filtration experiments were conducted with quartz prepared by a variety of other techniques. These methods are briefly described in Table II. Attachment efficiencies determined in these media are compared in Figure 7 with α values for the "well-cleaned" quartz. Colloids in "clean" media are more than an order of magnitude more mobile than colloids in

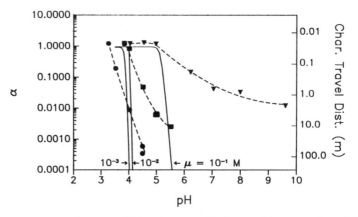

Figure 5.   Attachment efficiencies vs pH for 0.245 μm carboxyl latex spheres in well-cleaned quartz at varying ionic strengths: ($\blacktriangledown$) 0.1, ($\blacksquare$) 0.01, and ($\bullet$) 0.001 M.   Solid lines are theoretical predictions for the indicated ionic strength conditions.   The equivalent characteristic travel distance is plotted on the right y-axis.

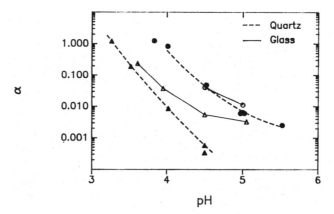

Figure 6.   Comparison of latex sphere attachment rates on well-cleaned quartz (ionic strengths: ($\blacktriangle$) = 0.001; ($\bullet$) = 0.01 M as NaCl) and glass beads (ionic strengths: ($\triangle$) = 0.001; ($\circ$) = 0.01 M as NaCl).

some of the dirtier systems. Only treatment methods "2" and "4" yielded comparable values of α as those obtained by the methods outlined above. The presence of organic impurities is implied by this result, and their removal requires either treatment with strong oxidants, such as chromic acid, or combustion at high temperatures. Other surface impurities, such as metals, are suspected since strong acid treatment was also necessary.

Attachment efficiencies of sulfate latex particles in glass bead media have been measured by other investigators and the cleaning methods used are summarized in Table II. Reported deposition rates in these filtration experiments, shown as points "5" and "6" in Figure 7, are slightly higher than the deposition rates of carboxyl latices in hydrochloric acid-washed quartz bed media and more than one order of magnitude greater than rates in well-cleaned media, even though the sulfate latex spheres have more negative zeta potentials at these pH conditions.

The manufacturing process of the quartz and glass materials may be the source of some of the apparent impurities. Our experience in sieving quartz, for example, indicates that mechanical abrasion of the sieves leads to contamination of the quartz surface, which we were unable to remove with concentrated HCl. Although various cleaning techniques, such as hydrofluoric acid treatment and calcining, are employed by suppliers, it is difficult to determine their efficiency, since only bulk concentrations of impurities are reported. After processing, the adsorption of impurities (particularly organic compounds) from the atmosphere is a likely source of additional contamination.

Conclusions

Comparisons of a variety of cleaning methods for silica-type collector surfaces indicate that high deposition rates of colloids are generally observed until aggressive measures are employed to remove impurities. While existing models that describe colloid-collector interactions in terms of DLVO theory still do not completely predict deposition rates in "clean" media, our results suggest a greater sensitivity of these rates to electrostatic repulsion than has been previously

Table II. Other Bed Media Cleaning Techniques Used
in Latex Sphere Filtration Experiments

Cleaning Method	Media	Reference
1. Conc. HCl soak	Quartz	This work
2. Conc. chromic acid soak; conc. HCl; boil in 4 N HNO₃	Quartz	This work
3. Conc. HCl soak; boil in 4 N HNO₃	Quartz	This work
4. Conc. HCl soak; heat (810° C, 8 hr)	Quartz	This work
5. 1 N HNO₃ soak; dry (60° C)	Glass beads	(6)[a]
6. Sonicate in 0.01 N NaOH; sonicate in 1 N HNO₃; dry (70-80° C)	Glass beads	(19)[a]

[a]Experiments were conducted with sulfate latex particles.

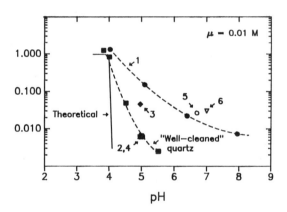

Figure 7. Comparison of latex sphere attachment efficiencies for quartz and glass beads with a variety of cleaning techniques. The number labels refer to the cleaning methods and materials listed in Table II. "Well-cleaned" quartz refers to the method outlined under "Experimental Procedures".

demonstrated. Experimental reproducibility also appeared to improve with the use of well-cleaned bed media. These findings underscore the inherent difficulty in understanding the surface charge characteristics of collector media at scales as small as the approaching colloid, on the basis of bulk measurements alone.

At low ionic strengths, and near the pH_{iep} of the latex spheres, filtration experiments provide additional supporting evidence for proposed "hairy-layer" models that describe the anomalous electrophoretic behavior of latex spheres. Using carboxyl latex spheres, apparent unfavorable filtration conditions can be observed at $pH < pH_{iep}$. Explanations for this finding can be qualitatively formulated if charge sites exist at the ends of flexible polymer chains.

For the conditions of our experiments, the corresponding characteristic travel distances of colloids were as high as approximately 100 m at low ionic strengths (see Figure 5). Although greater characteristic lengths may be possible, the design of experiments to answer this question is not trivial at laboratory scales. Given that the mobility of these colloids can be attributed to electrostatic repulsion, electrostatic forces may be useful in the development of manipulative strategies for colloids in aquifers.

Acknowledgments

Financial support from the U.S. Department of Energy (Grant DE-FG03-89ER60849) and a Clair A. Hill Scholarship from the Association of California Water Agencies is gratefully acknowledged.

Literature Cited

1.   Hull, M.; Kitchener, J. A. *Trans. Faraday Soc.* 1969, *65*, 3093.
2.   FitzPatrick, J. A.; Spielman, L. A. *J. Colloid Interface Sci.* 1973, *43*, 350-369.
3.   Rajagopalan, R.; Tien, C. *Can. J. Chem. Engr.* 1977, *55*, 256-264.
4.   Bowen, B. D.; Epstein, N. *J. Colloid Interface Sci.* 1979, *72*, 81-97.
5.   Gregory, J.; Wishart, A. *J. Colloids Surf.* 1980, *1*, 313-334.
6.   Elimelech, M.; O'Melia, C. R. *Environ. Sci. Technol.* 1990, *24*, 1528-1536.
7.   Li, H. C.; de Bruyn, P.L. *Surface Sci.* 1966, *5*, 203-220.
8.   Yao, K-M.; Habibian, M. T.; O'Melia, C. R. *Environ. Sci. Technol.* 1971, *5*, 1105-1112.
9.   Levich, V. *Physicochemical Hydrodynamics* 1962, Prentice Hall, Princeton, N.J.
10. Happel, J. *A. I. Ch. E. Journal* 1959, *5*, 174-177.
11. Spielman, L. A.; Friedlander, S. K. *J. Colloid Interface Sci.* 1974, *46*, 22-31.
12. Dahneke, B. *J. Colloid Interface Sci.* 1974, *48*, 520-522 .
13. Hogg, R.; Healy, T. W.; Fuerstenau, D. W. *Trans. Faraday Soc.* 1966, *62*, 1638-1651.
14. Gregory, J. *J. Colloid Interface Sci.* 1981, *83*, 138-145.
15. Zukoski, C. F.; Saville, D. A. *J. Colloid Interface Sci.* 1986, *114*, 45-53.
16. Van der Put, A. G.; Bijsterbosch, B. H. *J. Colloid Interface Sci.* 1983, *92*, 499-507.
17. Elimelech, M.; O'Melia, C. R. *Colloids Surfaces* 1990, *44*, 165-178.
18. Tien, C. *Granular Filtration of Aerosols and Hydrosols;* Butterworths: Boston, MA, 1989; p. 235-263.
19. Tobiason, J. E. *Colloids and Surfaces* 1989, *19*, 53-77.

RECEIVED December 11, 1991

Chapter 3

Deposition of Colloids in Porous Media
Theory and Numerical Solution

Menachem Elimelech and Lianfa Song

Civil and Environmental Engineering Department, School of Engineering
and Applied Science, University of California, Los Angeles, CA 90024

The convective diffusion equation describing the transport of suspended colloidal particles in the vicinity of stationary surfaces in
porous media was rigorously formulated by incorporating
fundamental theories of mass transport, hydrodynamics of particles
in porous media, and colloidal interactions. A finite difference
numerical scheme with a variable step size was developed to solve
the equation for the concentration distribution of colloidal particles
over solid surfaces in porous media, from which the colloid deposition
(capture) rates can be evaluated. With the mesh refinement technique
employed in the numerical scheme, no oscillations were observed in
the numerical solution.

Colloidal particles are ubiquitous in groundwater aquifers. The fate of many
pollutants in groundwater is determined in large part by the fate of the colloids
with which they are associated (1). Field studies indicate that colloids are mobile
in subsurface environments. Mobile colloids can act as a third phase that can
enhance transport of adsorbed pollutants. The enhancement in the transport of
pollutants sorbed to colloids is usually referred to as *facilitated transport* or
colloid-mediated transport. This process has not yet been considered in predictive models of contaminant transport.

Colloids interact with stationary surfaces in porous media. The extent of
migration of colloids and associated pollutants in groundwater is determined in
large part by the rate of their capture by aquifer stationary media. The capture
of colloidal particles from flowing suspensions by stationary surfaces is usually
referred to as particle deposition. In addition to the role of colloids in pollutant
transport, captured colloids can reduce the hydraulic conductivity of porous
media due to pore clogging (2,3). Artificial recharge of groundwater, for
instance, can be seriously hampered due to pore clogging caused by captured
colloidal materials (4). Control of pore clogging is also of considerable interest
in the field of petroleum extraction (3). An excellent review of the various
colloidal processes in groundwater and their significance was presented by
McDowell-Boyer et al. (1).

0097–6156/92/0491–0026$06.00/0

The literature already contains quantitative formulations of colloid deposition on various surfaces, including rotating disks, stagnation-point flow cells, and parallel-plate channels (5-7). These geometries are convenient for modeling, since the fluid flow field over these surfaces is simple and well defined. As a result, numerical solutions of the governing differential equations of these systems are straightforward. Furthermore, these simple surface geometries can easily be developed in the laboratory to test the governing equations.

The capture of colloids (submicrometer in size) in porous media, however, is more complex and has not been treated rigorously. This study presents a formulation and a numerical solution of the particle transport equation for colloids in porous media. The equation was developed by incorporating fundamental theories of mass transport, colloidal hydrodynamics, and colloidal interactions. In this model, it was assumed that the colloidal particles are spherical and that the porous medium is comprised of uniform spheres (referred to as collectors). A finite-difference numerical method using a mesh refinement technique is developed to solve the transport equation and to calculate the rate of colloid deposition in the porous medium.

Theory

The Particle Transport Equation. The transfer of colloidal particles from flowing suspensions toward stationary surfaces is governed by the convective diffusion equation. The equation, in its general form, is given by

$$\frac{\partial C}{\partial t} + \nabla \cdot \mathbf{J} = Q \tag{1}$$

where C is the concentration of particles, t is the time, $\mathbf{J}$ is the particle flux vector, and Q is a source/sink term. The particle flux vector is given by (8,9)

$$\mathbf{J} = -\mathbf{D} \cdot \nabla C + \mathbf{u}C + (1/kT)\,\mathbf{D} \cdot \mathbf{F}C \tag{2}$$

Here $\mathbf{D}$ is the particle diffusion tensor, $\mathbf{u}$ is the particle velocity vector induced by the flow of the suspending medium, k is the Boltzmann constant, T is the temperature, and $\mathbf{F}$ is the external force vector. The first, second, and third terms on the right hand side of equation 2 represent the transport of particles induced by diffusion, convection, and external forces, respectively. The external force vector can be derived from the interaction potential energy function, ϕ, as follows:

$$\mathbf{F} = -\nabla \phi \tag{3}$$

When equations 2 and 3 are substituted into equation 1, the convective diffusion equation under steady conditions and in the absence of a source term, is reduced to

$$\nabla \cdot (\mathbf{u}C) = \nabla \cdot (\mathbf{D} \cdot \nabla C) + \nabla \cdot \left(\frac{\mathbf{D} \cdot \nabla \phi}{kT} C \right) \tag{4}$$

Surface Interaction Potentials. The potential energy between suspended colloidal particles and collector surfaces resulting from surface interaction forces can be described by the Derjaguin-Landau-Verwey-Overbeek (DLVO) theory (8,10,11). In this approach, the total interaction potential is the sum of van der Waals and electric double layer interactions. Quantitative expressions for these interaction potentials, which were used in the theoretical calculations appearing in this work, are described below.

The universal van der Waals attraction forces have long been known to play a part in the capture of colloidal particles by surfaces. These forces arise from the fluctuating electromagnetic field between molecules of the colloids and collector surfaces (11). There are several analytical expressions which describe these surface forces between particles and stationary collectors. An excellent summary of these expressions was given by *Gregory* (12) and *Jia and Williams* (9).

In this work, the approximation of *Gregory* (12) for van der Waals attraction will be used. This expression is in good agreement with exact calculations at short distances (up to 20% of the particle radius) and is given by

$$\phi_{VDW} = - \frac{A a_p}{6y(1 + 14y/\lambda)} \tag{5}$$

where A is the Hamaker constant of the interacting media, a_p is the particle radius, y is the separation distance (surface to surface) between particles and collectors, and λ is the characteristic wavelength of the interaction, often assumed to be 100 nm. For larger separations, the exact expression for retarded van der Waals interaction derived by *Czarnecki* (13) will be used. This expression is given by

$$\phi_{VDW} = -A \left\{ \frac{2.45\lambda}{60\pi} \left(\frac{y - a_p}{y^2} - \frac{y + 3a_p}{(y + 2a_p)^2} \right) - \frac{2.17\lambda^2}{720\pi^2} \left(\frac{y - 2a_p}{y^3} - \frac{y + 4a_p}{(y + 2a_p)^3} \right) \right.$$
$$\left. + \frac{0.59\lambda^3}{5040\pi^3} \left(\frac{y - 3a_p}{y^4} - \frac{y + 5a_p}{(y + 2a_p)^4} \right) \right\} \tag{6}$$

Czarnecki's expression is inaccurate for distances shorter than about 10 nm; at such short range, however, equation 5 is adequate.

Most surfaces and colloids in aqueous media are charged and have electric double layers associated with them (10,11). The surface charge is balanced by an equivalent number of counterions, some of which are located very close to the surface, in the so-called *Stern layer*, while the remainder are distributed in the *diffuse layer*. When a charged particle approaches a similarly charged surface, their diffuse layers overlap, and, as a result, a repulsive force develops.

Analytical expressions for this force as a function of the separation distance are available in the literature (*10,14-16*). The widely used analytical expression of *Hogg et al.* (*14*) for electrical double layer interaction will be used in this study. It gives the potential energy as

$$\phi_{EDL} = \pi\varepsilon_0\varepsilon_r a_p \left\{ 2\psi_1\psi_2 \ln\left(\frac{1+\exp(-\kappa y)}{1-\exp(-\kappa y)}\right) + (\psi_1^2 + \psi_2^2) \ln\left[1-\exp(-2\kappa y)\right] \right\} \quad (7)$$

Here ε_r and ε_0 are the relative dielectric permittivity of water and the permittivity under vacuum, respectively; ψ_1 and ψ_2 are the surface potentials of particles and collectors, respectively; and κ is the reciprocal Debye length. This expression is valid for: (i) interaction at constant potential (ii) 1:1 electrolytes; and (iii) surface potentials smaller (in absolute value) than 60 mV.

Hydrodynamics of Particles in Porous Media. The role of hydrodynamics in the transport of colloids to stationary surfaces depends upon the geometry of the collectors, the flow field, and the physical properties of the particles and the liquid phase. This study is concerned with the deposition of submicron spherical particles on stationary surfaces in porous media at low Reynolds number flow. This condition (i.e., low Reynolds number flow) is typical of groundwater flow.

Several models describing the flow field in porous media at low Reynolds numbers are available (*17-19*). Among these, Happel's sphere-in-cell flow model (*17*) is the most commonly used. Happel's model has also been used successfully in particle filtration models in which the transfer of particles to stationary collectors is involved. In Happel's model, the porous medium is treated as an assemblage of identical spherical collectors, each of which is enveloped in a shell of fluid. The thickness of the shell is determined so that the overall porosity of the porous medium is maintained for the single collector (*17,20*).

The undisturbed fluid flow field in Happel's model is obtained from a solution of the Navier-Stokes equation. The stream function for axisymmetric, steady creeping flow around a spherical collector in this model is given by (*19*)

$$\Psi = \frac{1}{2} U a_c^2 \sin^2\theta \left[K_1\left(\frac{a_c}{r}\right) + K_2\left(\frac{r}{a_c}\right) + K_3\left(\frac{r}{a_c}\right)^2 + K_4\left(\frac{r}{a_c}\right)^4 \right] \quad (8)$$

where a_c is the radius of the collector; U is the approach velocity of the fluid; r is a radial coordinate originating from the center of the spherical collector; and K_1, K_2, K_3 and K_4 are constants that depend on the porosity of the porous medium. The radial and tangential components of the fluid velocity around the spherical collector (v_r and v_θ, respectively) can be derived from equation 8 as follows:

$$v_\theta = \frac{1}{r\sin\theta}\frac{\partial\Psi}{\partial r} \quad (9)$$

$$v_r = -\frac{1}{r^2\sin\theta}\frac{\partial\Psi}{\partial\theta} \quad (10)$$

where r and θ are the radial and tangential coordinates of the spherical coordinate system used in this work.

Even in the absence of surface interaction forces, the trajectories of particles at short distances from collector surfaces do not follow the fluid streamlines. This deviation is caused by hydrodynamic (viscous) interactions (21). In a viscous fluid, such as water, the approach of a particle to a surface is hindered by the slow drainage of water from the narrowing gap between the particle and surface. As the gap narrows to zero, this drainage becomes infinitely slow because of the no-slip conditions at the surfaces and, as a result, the drag on the particles significantly increases. Considering hydrodynamic interaction, the radial and tangential particle velocity components (u_r and u_θ, respectively) are related to the fluid velocity components by (6,22)

$$u_r = f_1(H)\, f_2(H)\, v_r \tag{11}$$

$$u_\theta = f_3(H)\, v_\theta \tag{12}$$

Here $f_1(H)$, $f_2(H)$, and $f_3(H)$ are universal correction factors for hydrodynamic interaction; these depend on the dimensionless distance ($H = y/a_p$) between the particle and the surface of collectors (21, and references therein).

The decrease in the mobility of particles in the vicinity of the collector surface due to hydrodynamic interactions results in a decrease in their diffusion coefficient. In this case, the diffusion coefficients are expressed as (9)

$$D_r = f_1(H)\, D_\infty \tag{13}$$

$$D_\theta = f_4(H)\, D_\infty \tag{14}$$

where $f_4(H)$ is a hydrodynamic correction factor, and D_∞ is the particle diffusion coefficient at large distances from the collector. The latter is a scalar and can be obtained from the Stokes-Einstein equation (19):

$$D_\infty = \frac{kT}{6\pi\mu a_p} \tag{15}$$

Mathematical Formulation

Particle Transport Equation in Spherical Coordinates. For a porous medium comprised of spherical collectors, the dimensionless convective diffusion equation in spherical coordinates can be written as

$$\frac{N_{P_e}}{2}\left(f_1(H)\, f_2(H)\, V_r \frac{\partial C^*}{\partial H} + \frac{f_3(H) V_\theta}{N_R + H}\frac{\partial C^*}{\partial \theta} + f_1(H)\, f_2(H)\, C^* \frac{\partial V_r}{\partial H} + f_1(H)\, V_r C^* \frac{\partial f_2(H)}{\partial H} + f_2(H)\, V_r C^* \frac{\partial f_1(H)}{\partial H} \right.$$

$$\left. + \frac{2 f_1(H)\, f_2(H)}{N_R + H} V_r C^* + \frac{\cot\theta}{N_R + H} 2 f_3(H)\, V_\theta C^* \right) = \left(\frac{\partial f_1(H)}{\partial H}\frac{\partial C^*}{\partial H} + \frac{2 f_1(H)}{N_R + H}\frac{\partial C^*}{\partial H} + f_1(H)\frac{\partial^2 C^*}{\partial H^2} \right.$$

$$\left. + \frac{2 f_1(H)}{N_R + H} C^* \frac{\partial \Phi}{\partial H} + C^* \frac{\partial f_1(H)}{\partial H}\frac{\partial \Phi}{\partial H} + f_1(H)\frac{\partial C^*}{\partial H}\frac{\partial \Phi}{\partial H} + f_1(H)\, C^* \frac{\partial^2 \Phi}{\partial H^2} \right) \tag{16}$$

Here $C^* = C/C_0$, C_0 being the concentration of particles in the approaching fluid; H is a dimensionless separation distance from the collector surface (defined previously); $\Phi = \phi/kT$ is a dimensionless potential energy, ϕ being the total interaction energy between particles and collector; $V_r = v_r/U$ and $V_\theta = v_\theta/U$, v_r and v_θ being the radial and tangential velocity components of the fluid defined by equations 9 and 10; $N_{Pe} = (2a_p U)/D_\infty$ is a particle Peclet number; and $N_R = (a_c + a_p)/a_p$. In the derivation of this equation from equation 4, the following assumptions were made: (i) colloidal forces act perpendicular to the surfaces; and (ii) particle diffusive fluxes tangent to the collector surface are much smaller than the radial (perpendicular) diffusive fluxes.

For convenience in numerical calculations, equation 16 can be rewritten as

$$\frac{\partial C^*}{\partial \theta} = a_1(H, \theta) \frac{\partial^2 C^*}{\partial H^2} + a_2(H, \theta) \frac{\partial C^*}{\partial H} + a_3(H, \theta) C^* \tag{17}$$

where the functions $a_1(H, \theta)$, $a_2(H, \theta)$, and $a_3(H, \theta)$ can be determined from equation 16. The boundary conditions in our case are

$$C^*(H = 0, \theta) = 0 \tag{18a}$$

$$C^*(H \to \infty, \theta) = 1 \tag{18b}$$

$$\left(\frac{\partial C^*}{\partial \theta} \right)_{\theta = 0} = 0 \tag{18c}$$

In the first boundary condition, the concentration of particles in contact with the collector is taken to be zero because these particles are no longer part of the dispersed phase. The deposition at $H=0$ is assumed to be irreversible (the so-called "perfect sink" model). The second boundary condition states that at large separations from the surface of the collectors, the concentration of particles is equal to that of the approaching fluid. In Happel's porous medium model, this boundary condition can be taken at the outer surface of the fluid envelope; the thickness of the fluid envelope is determined by the porosity and collector diameter (*19*). The third boundary condition arises from the symmetry around the forward stagnation path of the spherical collector (*22*).

Calculation of Particle Deposition Rates. Once the dimensionless concentration distribution of particles around the collector, $C^*(H, \theta)$, is determined, the perpendicular flux of particles at the collector surface can be calculated. When the local particle flux at the surface (i.e., at $H=0$) is integrated over the entire surface of the collector, the overall rate of particle deposition is obtained. The local particle flux perpendicular to the surface can be expressed as

$$J(H, \theta) = UC_0 J^*(H, \theta) \tag{19}$$

where $J^*(H, \theta)$ is a dimensionless flux given by

$$J^*(H, \theta) = -\frac{2f_1(H)}{N_{P_e}} \frac{\partial C^*}{\partial H} + f_1(H)f_2(H)V_r C^* - \frac{2f_1(H)}{N_{P_e}} \frac{\partial \Phi}{\partial H} C^* \qquad (20)$$

In this equation, the first term describes the diffusive flux of particles, the second term describes the convective flux of particles, and the third term describes the flux of particles due to migration (drift) velocity resulting from the interaction potentials (i.e., van der Waals and electrical double layer). The dimensionless local flux of particles at the collector surface is evaluated from equation 20 with $H \rightarrow 0$.

The overall rate of particle deposition on the collector, I, can be obtained from integration of the local flux (at $H=0$) over the entire surface of the collector as follows:

$$I = 2\pi a_c^2 \int_0^\pi J(H = 0, \theta) \sin\theta d\theta \qquad (21)$$

In studies concerned with deposition of colloidal particles in porous media, it is convenient to use a dimensionless deposition rate, also known as the single collector efficiency,

$$\eta = \frac{I}{\pi a_c^2 U C_0} \qquad (22)$$

This can be viewed as the ratio of the overall deposition rate of particles on the collector to the convective transport of upstream particles towards the projected area of the collector.

Numerical Solution

General Considerations. Equation 17 is similar in form to the one-dimensional advection-dispersion equation with a first-order chemical reaction:

$$\frac{\partial C}{\partial t} = D_d \frac{\partial^2 C}{\partial x^2} - v_x \frac{\partial C}{\partial x} - k_r C \qquad (23)$$

Here D_d is the dispersion coefficient, v_x is the average fluid velocity in the direction x, and k_r is the reaction rate constant. The advection-dispersion equation is used to describe the mass transport of solutes in various processes. The features of the equation have been the subject of numerous reviews (23,24). In the numerical solution of equation 17, some concepts and results will be adapted from previous studies of the advection-dispersion equation.

When the ratio of advection (the second term in the right hand side of equation 23) to dispersion (the first term in the right hand side of equation 23) is small to moderate (so-called "dispersion-dominated transport"), no numerical difficulties in the solution of this equation are encountered. Several numerical methods to solve this equation are available for this case (*23,24*). On the other hand, when this ratio is large (so-called "advection-dominated transport"), severe numerical difficulties are encountered when these methods are employed. These numerical difficulties are of two types: overshoot and numerical dispersion (*23,25*). A mesh Peclet number is usually used in numerical calculations to characterize the ratio of advection to dispersion. The mesh Peclet number is defined as $v_x \Delta x / D_d$, Δx being the step size in the direction x. Most numerical methods give an accurate solution when the mesh Peclet number is smaller than unity. However, when the mesh Peclet number increases, oscillations appear in the numerical solution.

In the problem of capture of colloidal particles by stationary surfaces, the interaction potential near the surface is extremely large due to the nature of the universal van der Waals attraction forces. Consequently, the velocity of the particles at the collector surface becomes infinitely large. In addition, the diffusion coefficient near the surface approaches a very small value due to the retardation of particle mobilities caused by hydrodynamic interactions. As a result, at the collector surface, the coefficient $a_1(H, \theta)$ in equation 17 approaches zero while $a_2(H, \theta)$ approaches infinity. Thus, close to the collector surface equation 17 behaves as the advection-dispersion equation with high Peclet number; oscillations in the numerical solution of equation 17 are obtained in this region when conventional methods are used. These oscillations affect the accuracy of the numerical solution for the concentration distribution of particles, and consequently for the particle deposition rates in the porous medium. In order to overcome the numerical difficulties in the solution of equation 17, a particular mesh refinement technique near the collector surface is used in the numerical scheme.

Numerical Discretization. Equation 17 with the boundary conditions described by equations 18a-18c will be solved in two stages. In the first stage, by applying boundary condition 18c, the equation is reduced to an ordinary differential equation:

$$a_1(H,0) \frac{d^2 C^*}{dH^2} + a_2(H,0) \frac{dC^*}{dH} + a_3(H,0)\, C^* = 0 \qquad (24)$$

Equation 24 with the boundary conditions 18a and 18b is a typical stiff two-point boundary value problem. A central difference scheme on a non-uniform mesh, developed by Pearson (*26*), will be used to solve this problem. The solution yields the concentration distribution of particles at the forward stagnation point of the spherical collector (i.e., at $\theta = 0$). In the second stage, the concentration distribution of particles over the entire region of the spherical collector will be calculated (i.e., for $0 < \theta < \pi$). In this stage, the concentration distribution of particles at $\theta = 0$ (from the first stage) will be used as a boundary condition in addition to equations 18a and 18b. A modified Crank-Nicolson scheme with a non-uniform mesh is employed in the numerical solution of this stage.

With the central difference scheme on a non-uniform step size, the first and second derivatives of equation 24 can be approximated by

$$\left(\frac{dC^*}{dH}\right)_i = \frac{q_i}{p_i+q_i}\frac{C^*_{i+1}-C^*_i}{p_i} + \frac{p_i}{p_i+q_i}\frac{C^*_i-C^*_{i-1}}{q_i} \tag{25}$$

$$\left(\frac{d^2C^*}{dH^2}\right)_i = \frac{2}{p_i+q_i}\left(\frac{C^*_{i+1}-C^*_i}{p_i} - \frac{C^*_i-C^*_{i-1}}{q_i}\right) \tag{26}$$

where i is the index of the space variable H; C^*_i is the dimensionless concentration at point H_i; $p_i = H_{i+1} - H_i$; and $q_i = H_i - H_{i-1}$. From equations 25 and 26, equation 24 can be approximated by

$$e_{1i}C^*_{i+1} + e_{2i}C^*_i + e_{3i}C^*_{i-1} = 0 \tag{27}$$

where e_{1i}, e_{2i}, and e_{3i} are variable coefficients (depending on H) that can be determined from equations 24-26. This equation is readily solved by the Gauss elimination method (Thomas algorithm).

In the solution of equation 17 for the entire region (i.e., for $\theta > 0$), the conventional Crank-Nicolson method is inadequate, due to oscillations in the numerical solution. As discussed previously, these oscillations can be attributed to the extremely large particle velocities in the vicinity of the collector surface. In order to avoid such oscillations, the Crank-Nicolson scheme is modified to include a non-uniform mesh. A variable step size in direction H will be employed to eliminate these oscillations (see below).

The finite difference approximations for the various derivatives in equation 17 are as follows:

$$\left(\frac{\partial C^*}{\partial \theta}\right)_{n+1/2,i} = \frac{1}{\Delta\theta}(C^{*n+1}_i - C^{*n}_i) \tag{28}$$

$$\left(\frac{\partial C^*}{\partial H}\right)_{n+1/2,i} = \frac{1}{2}\left\{\left[\frac{q_i}{p_i+q_i}\frac{C^{*n+1}_{i+1}-C^{*n+1}_i}{p_i} + \frac{p_i}{p_i+q_i}\frac{C^{*n+1}_i-C^{*n+1}_{i-1}}{q_i}\right]\right.$$
$$\left. + \left[\frac{q_i}{p_i+q_i}\frac{C^{*n}_{i+1}-C^{*n}_i}{p_i} + \frac{p_i}{p_i+q_i}\frac{C^{*n}_i-C^{*n}_{i-1}}{q_i}\right]\right\} \tag{29}$$

$$\left(\frac{\partial^2 C^*}{\partial H^2}\right)_{n+1/2,i} = \frac{1}{2}\left\{\frac{2}{p_i+q_i}\left[\frac{C^{*n+1}_{i+1}-C^{*n+1}_i}{p_i} - \frac{C^{*n+1}_i-C^{*n+1}_{i-1}}{q_i}\right]\right.$$
$$\left. + \frac{2}{p_i+q_i}\left[\frac{C^{*n}_{i+1}-C^{*n}_i}{p_i} - \frac{C^{*n}_i-C^{*n}_{i-1}}{q_i}\right]\right\} \tag{30}$$

where n is the index of the tangential coordinate θ. Substituting equations 28-30 in equation 17 and rearranging, we obtain

$$b_{1i}^{n+1/2} C_{i-1}^{*\,n+1} + b_{2i}^{n+1/2} C_i^{*\,n+1} + b_{3i}^{n+1/2} C_{i+1}^{*\,n+1}$$

$$= d_{1i}^{n+1/2} C_{i-1}^{*\,n} + d_{2i}^{n+1/2} C_i^{*\,n} + d_{3i}^{n+1/2} C_{i+1}^{*\,n} \tag{31}$$

where the coefficients $b_{1i}^{n+1/2}, b_{2i}^{n+1/2}, b_{3i}^{n+1/2}, d_{1i}^{n+1/2}, d_{2i}^{n+1/2}$, and $d_{3i}^{n+1/2}$ can be obtained from equations 17, and 28-30.

Mesh Refinement Technique. Due to the extremely high particle velocities near the collector surface, the perpendicular coordinate (H) was divided into two regions, an inner and outer region. In the inner region, where the interactions are very strong, an extremely small step size is required in order to prevent oscillations in the numerical solution. An exponentially decreasing step size (decreasing towards the surface of the collector) was employed in this study. The varying step size was calculated from

$$H_i = \delta + (L_1 - \delta) \frac{\exp(i\lambda/N_1) - 1}{\exp(\lambda) - 1} \tag{32}$$

where L_1 is the length of the inner region; δ is the first point in H (for $i=0$); N_1 is the number of steps in the inner region; and λ is a positive number adjusting the rate of change of the step size. With this equation, extremely small step sizes (on the order of 10^{-12}m) were generated in the vicinity of the collector. The step size increases gradually towards the end of the inner region. In the outer region where the interaction forces can be neglected, a uniform step size with relatively large ΔH was employed.

Based on boundary condition 18a, the first point (δ) is zero (perfect sink model). An arbitrary value of $\delta = 2 \times 10^{-5}$ was selected; this can be assumed as practically zero since for submicron particles it is equivalent to particle-collector separation distances smaller than the size of an atom. The calculated particle deposition rates varied by less than 1% when δ was increased by more than one order of magnitude. The outer boundary condition, 18b, was calculated from Happel's model; this is the radius of the fluid envelope of Happel's model. For the numerical simulations conducted in this study, the outer boundary condition was applied at separation distances close to 10,000 nm.

Preliminary calculations were carried out in order to determine the size of the inner and outer regions. These calculations were carried out for particle deposition in the presence of attractive double layer interactions at low ionic strengths. Under such conditions, electrical double layer attraction forces are very strong, and, as a result, the particle velocity approaches an extremely large value in the vicinity of the collector surface. Calculations showed that when the inner region was larger than 0.5% of the entire range, particle deposition rates approached a constant value. At smaller inner regions (i.e., less than 0.5%), oscillations appeared in the numerical solution. This is due to the fact that part of the range where attractive forces are large was in the outer region. Based on these results, the inner region in the calculations presented in this paper was taken as 1% of the entire range.

Additional calculations demonstrated that when the number of steps in the inner region (N_1) exceeded 300, there were no oscillations in the numerical solution. Accordingly, it was decided to use $N_1 = 500$. Furthermore, the optimal number of steps in the outer region was found to be 2500. With the number of steps in the outer region smaller than this value, the calculated deposition rates varied slightly as the number of steps was changed. This is most probably due to truncation error.

The efficiency of the mesh refinement technique used in this study is illustrated in Figure 1. In this figure, the calculated particle deposition rates at the collector surface and at various spherical surfaces concentric with the collector ($H>0$) are presented. The latter were obtained by integration of the local fluxes at different H values. It should be emphasized that actual particle deposition occurs only at the collector surface (i.e., at $H=0$). The calculations for $H>0$ are only to demonstrate the efficiency of the mesh refinement technique. The displayed deposition rates were calculated with the mesh refinement technique (exponential step size) and with uniform step size for the case of attractive double layer interactions. With the exponential step size, 500 steps were used in the inner region and 2500 in the outer region. As can be seen from Figure 1, the particle deposition rates calculated with the exponential step size were constant over the displayed range. This is expected for particle deposition onto a spherical collector in the absence of repulsive double layer interactions

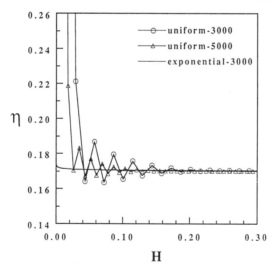

FIGURE 1: Overall deposition rates of particles at the collector surface ($H=0$) and at various concentric surfaces ($H>0$) when double layer interactions are attractive. The deposition rates were numerically calculated using uniform and exponential step sizes. The following parameters were used in the calculations: Hamaker constant = 5×10^{-20} J; particle diameter = 0.4 μm; collector diameter = 0.1 mm; porosity = 0.38; approach velocity = 10^{-5} m/s; surface potential of particles and collectors = 20 and -20 mV, respectively; ionic strength = 10^{-5} M; and temperature = 25°C.

(22). In the calculations with the uniform step size, on the other hand, oscillations in the numerical solution were obtained in the vicinity of the collector surface due to the large mesh Peclet number. As a result, particle deposition rates near the collector surface deviated significantly from those calculated with the exponential step size. At larger distances from the collector, the mesh Peclet number decreases, and particle deposition rates approach those calculated with the exponential step size.

The superiority of the mesh refinement technique is also illustrated in Figure 2. In this figure, particle deposition rates at the collector surface (i.e. at $H=0$), calculated with uniform step size, are described as a function of the number of steps (in H) used in the numerical solution. As the number of steps increases, particle deposition rates become closer to those calculated with the mesh refinement technique. However, calculations similar to those presented in Figure 1 showed that oscillations at the collector surface do not disappear entirely even when the number of steps is 50,000. As a result, particle deposition rates calculated at the collector surfaces deviate from the true value (i.e., that calculated with the exponential step size). When the mesh refinement technique is used, only 3000 steps are required in order to eliminate oscillations.

Comparison with a Limiting Analytical Solution. Equation 17 with the boundary conditions 18a-18c has no analytical solution. However, under several simplifying assumptions, an approximate analytical solution for the particle deposition rate is available. A comparison of the numerical solution derived in this work with a limiting analytical solution can provide a test for the accuracy of the numerical scheme.

When colloidal interactions are not considered (i.e., $\phi = 0$) and hydrodynamic interactions are ignored (i.e., $f_i(H) = 1$ for $i = 1$-4), the convective diffusion equation can be solved analytically. The particle deposition rate with Happel's porous medium model (17) under these conditions is given analytically by (27)

$$\eta = 4.0\, A_S^{1/3} \left(\frac{D_\infty}{U d_c} \right)^{2/3} \tag{33}$$

Here d_c is the collector diameter and A_S is a porosity dependent parameter given by

$$A_S = \frac{2(1 - \beta^5)}{2 - 3\beta - 3\beta^5 - 2\beta^6} \tag{34}$$

where $\beta = (1 - f)^{1/3}$, f being the porosity of the porous medium; and D_∞ is the diffusion coefficient at infinite separation (equation 15). Calculations over a wide range of particle sizes (smaller than $0.5\mu m$) and fluid velocities showed that particle deposition rates calculated from equation 33 and from a numerical solution of the convective diffusion equation (under the assumptions discussed above) differed by less than 2%.

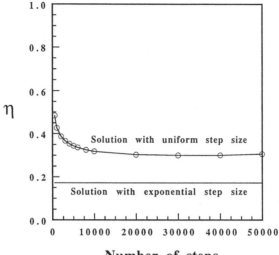

η

Number of steps

FIGURE 2: Deposition rates of particles at the collector surface as a function of the number of steps used in the numerical solution with uniform step size. The capture rate calculated with the mesh refinement technique (3000 steps) is also presented. The physical and chemical parameters were similar to those used in Figure 1.

Summary and Conclusions

A quantitative formulation of the convective diffusion equation for transport of colloidal particles in porous media was presented. The formulation of this equation was based on fundamental theories of mass transport, colloidal hydrodynamics, and colloidal interactions. A unique feature of the convective diffusion equation is the dependence of the particle Peclet number on the distance of the colloidal particles from the surface of the stationary collectors of the porous medium. At the vicinity of the collector surface, the particle Peclet number is extremely large (convection-dominated transport), while far from the collector surface, the particle Peclet number is very small (diffusion-dominated transport).

A finite-difference numerical scheme to solve this equation was developed. In this numerical scheme, a mesh refinement technique was used to overcome the numerical difficulties in the vicinity of the collector surface. The step size close to the collector surface was varied exponentially so that the mesh Peclet number was kept relatively small for all separation distances. With this method, there were no oscillations in the numerical solution. A total of 3,000 steps was adequate to eliminate oscillations when the mesh refinement technique was used.

Acknowledgements

The authors acknowledge the financial support of the National Science Foundation (Research Grant BCS-9009233) and the UCLA Academic Senate.

Literature Cited

1. McDowell-Boyer, L.M.; Hunt, J.R.; Sitar, N. *Water Resour. Res.*, 1986, 22, 1901-1921.
2. Goss, D.W.; Smith, S.J.; Stuart, B.A.; Jones, O.R. *Water Resour. Res.*, 1976, 9, 668-675.
3. Sahimi, M.; Gavalas, G.R.; Tostis, T.T. *Chem. Eng. Sci.*, 1990, 45, 1443-1502.
4. *Artificial Recharge of Groundwater*, Asano, T., Ed., Butterworths: Stoneham, MA; 1985.
5. Bowen, B.D.; Levin, S.; Epstein, N. *J. Colloid Interface Sci.*, 1976, 54, 375-390.
6. Adamczyk, Z.; Czarnecki, J.; Dabros, T.; Van de Ven, T.G.M. *Adv. Colloid Interface Sci.*, 1983, 19, 183-252.
7. Chari, K.; Rajagopalan, R. *J. Chem. Soc. Faraday Trans.*, 1985, 81, 1345-1366.
8. Van de Ven, T.G.M. *Colloidal Hydrodynamics*; Academic Press, London, 1989.
9. Jia, X.; Williams, R.A. *Chem. Eng. Comm.*, 1990, 91, 127-198.
10. Verwey, E.J.W.; Overbeek, J. Th. G. *Theory of the Stability of Lyophobic Colloids*; Elsevier, Amsterdam, 1948.
11. Israelachvili, J.N. *Intermolecular and Surface Forces*; Academic Press, London, 1985.
12. Gregory, J. *J. Colloid Interface Sci.*, 1981, 83, 138-145.
13. Czarnecki, J. *J. Colloid Interface Sci.*, 1979, 72, 361-362.
14. Hogg, R.; Healy, T.W.; Fuerstenau, D.W. *Trans. Faraday Soc.*, 1966, 66, 1638-1651.
15. Wiese, G.R.; Healy, T.W. *Trans. Faraday Soc.*, 1970, 66, 490-499.
16. Gregory, J. *J. Colloid Interface Sci.*, 1975, 61, 44-51.
17. Happel, J. *AIChE J.*, 1958, 4, 197-201.
18. Kuwabara, S. *J. Phys. Soc. Japan*, 1959, 14, 527-532.
19. Tien, C. *Granular Filtration of Aerosols and Hydrosols*; Butterworths Publishers, Stoneham, MA, 1989.
20. Pfeffer, R.; Happel, J. *AIChE J.*, 1964, 10, 605-611.
21. Spielman, L.A. *Annu. Rev. Fluid Mech.*, 1977, 9, 297-319.
22. Prieve, D.C.; Ruckenstein, E. *AIChE J.*, 1974, 20, 1178-1187.
23. Huyakorn, P.S.; Pinder, G.F. *Computational Methods in Subsurface Flow*; Academic Press, London, 1983.
24. Sun, N-Z. *Environ. Health Prosp.*, 1989, 83, 97-115.
25. Gray, W.G.; Pinder, G.F. *Water Resour. Res.*, 1976, 12, 547-555.
26. Pearson, C.E. *J. Math. Phys.*, 1968, 47, 134-154.
27. Spielman, L.A.; Friedlander, S.K. *J. Colloid Interface Sci.*, 1974, 46, 22-31.

RECEIVED December 11, 1991

Chapter 4

Surface-Charge Repulsive Effects on the Mobility of Inorganic Colloids in Subsurface Systems

Robert W. Puls[1] and Robert M. Powell[2]

[1]Robert S. Kerr Environmental Research Laboratory, U.S. Environmental Protection Agency, Ada, OK 74820
[2]ManTech Environmental Technology, Inc., Ada, OK 74820

Batch and column experiments using natural aquifer material investigated the specific adsorption of anions onto charged inorganic colloidal surfaces in terms of enhanced colloid stability and transport in subsurface model systems. Variables in the study included flow rate, pH, ionic strength, aqueous chemical composition, colloid concentration and size. Specific adsorption of some anions resulted in enhanced colloid stability and transport of Fe_2O_3 particles due to increases in charge repulsion between the particles in suspension and between the particles and the immobile column matrix minerals. Extent of particle breakthrough was dependent upon a complex variety of parameters; however, the highest statistical correlation was observed with particle size and ionic composition of the supporting electrolyte.

The hydrogeochemical significance of colloidal-size particles in subsurface systems has only been realized during the past few years. This realization has resulted from field studies that show contaminant migration over distances and at concentrations greater than model predictions would allow. These models generally perform predictive calculations by assuming the contaminants interact with the mobile aqueous and immobile solid phases, and occasionally include the possibilities of contaminant free-phase and co-solvency effects. They account for solubility, speciation, ion-exchange, adsorption-desorption and diffusion reactions within and between these phases but do not consider the possibility of these interactions with a potential additional phase, mobile colloidal solids. Should such a phase be present in sufficient quantity, exhibit high sorption reactivity, remain stable in suspension, and be capable of avoiding attachment to the immobile solid phase, it might serve as an important mechanism for contaminant transport.

Colloids are particles that are sufficiently small that the surface free

0097–6156/92/0491–0040$06.00/0

energy of the particle dominates the bulk free energy. Typically this includes particles with diameters between 1 and 1000 nm (1,2). Colloidal particles can be organic, inorganic, or a combination of the two. Although the term "colloid" has classically referred to the system of the dispersed phase (e.g., Fe_2O_3) in the dispersing medium (e.g., NaCl), many references to "colloid" have altered the definition and often include only the particles themselves. Due to their small size, these particles typically have a large surface area per unit mass. If this surface contains functional groups or hydrophobic moieties, the particle can become a significant sink for ionic (both inorganic and organic) or neutral organic contaminants. The equilibrium sorption of the contaminant per unit of sorbent colloid mass can be much higher than would be estimated by sorption studies on larger particles of the same material, due to the surface area to volume effect.

Colloids do exist in ground-water systems at significant concentrations. Estimates of suspended particles in several ground-water systems have ranged as high as 63 mg/L (2-4). Several mechanisms can account for the presence of suspended stable particles in ground water, including:

1) dissolution of the soil or mineral matrix cement due to changes in pH or redox conditions,

2) supersaturation of the system with respect to an inorganic species that results in the formation, by nucleation and precipitation, of an inorganic colloid,

3) physical disruption of the mineral matrix by large alterations in flow conditions due to contaminant injection, ground-water withdrawal, large rainfall infiltrations, tidal influences, etc.,

4) the relatively slow, natural release of particles due to matrix dissolution and weathering of clays,

5) the release and movement of viruses and bacteria,

6) emulsions or microemulsions formed as the result of mixed waste solvents,

7) macromolecules and micelles formed from the agglomeration of humic acid molecules or surfactants.

It is important to note that the arrival of a contaminant plume can result in the formation of colloidal particles through the first three processes listed. A subsequent reduction in ionic strength, due for example to infiltration of lower ionic strength water, can enhance the stability of the particles and increase their transportability.

Numerous studies have demonstrated the reactivity of these ubiquitous ground-water particles. Secondary clay minerals, hydrous iron, aluminum, and manganese oxides, and humic materials have all been shown to be strongly adsorptive (5-8). Organic carbon colloids were found to be the major factor controlling the distribution of plutonium between the solid and aqueous phases by Nelson et al. (9). Buddemeier and Rego (10) found the activity of Mn, Co, Sb, Cs, Ce and Eu to be primarily associated with colloidal particles in samples from the Nevada Test Site. In laboratory column studies, Sandhu and Mills (7) determined that more than 90% of the chromium and arsenic present were sorbed to colloidal iron and manganese oxides.

These reactive particles have also been shown to be mobile under a variety of conditions in both field studies and laboratory column experiments (11-16). Size has been shown to be a significant parameter in colloid mobility and transport. Reynolds (17) studied the transport of carboxylated polystyrene beads in laboratory sand columns and found column effluent recoveries to be lower for the larger diameter particles (45% at 910 nm) and higher for the smaller particles (70% at 100 and 280 nm). Size has also been shown to affect transport time, with larger particles generally having faster transport than smaller particles (18,19) and, in some cases, faster than a conservative dissolved tracer such as chloride, bromide or tritiated water. This is due to the principle of size exclusion, known from the column chromatography literature. The larger particles are physically excluded from passage through the smaller pore spaces in the porous media due to their size, resulting in a reduced path length relative to the dissolved solutes.

Solution chemistry is extremely important in terms of colloid stability, mobility, and reactivity due to its effect on surface charge phenomena. Inorganic particles generally carry a charge that is either net negative or net positive depending on a number of factors. These factors include mineralogy (i.e. crystalline structure), solution pH, solution ionic strength, and the presence or absence of strongly adsorbing potential determining ions. Mineral species can possess either a fixed surface charge, a variable surface charge, or a combination of the two. In general, the immobile aquifer material will possess a net negative charge due to the preponderance of silica in the matrix. If the solution chemistry is such that the colloids are positively charged, then the particles can be attracted by and attached to immobile surfaces. Solution chemistry can also affect the particle-particle interactions such as attraction, that results in agglomeration and settling, or repulsion that can maintain particles suspended in the mobile fluid phase. Matijevic et al. (20) found that deposition and detachment of hematite particles in columns packed with stainless-steel beads were dominated by surface charge effects due to differences in electrolyte composition. In simulated aquifer experiments using sand beds, Champlin and Eichholz (21) demonstrated that previously "fixed" particles and associated contaminants may be remobilized by changes in the aqueous geochemistry of the system.

The purpose of this research was to evaluate specific aqueous chemical effects on the stability and transport of inorganic particles through natural porous media. Repeated particle size analysis of suspended particles at several field sites (including the Globe, AZ site) have indicated a preponderance of colloids in the size range 100 to 2000 nm present in samples collected at very low pumping rates. Because of these and similar observations by Gschwend and Reynolds (11), spherical iron oxide particles 100 to 900 nm in diameter were synthesized for use as the mobile colloidal phase in laboratory column experiments. Alluvium from the Globe site was used as the immobile column packing material. The extent of particle transport was investigated as a function of changes in column flow rate, pH, ionic strength, electrolyte composition, particle concentration and size. Batch experiments were performed to evaluate colloid stability.

Experimental Methods

Characterization of Aquifer Solids. Column matrix material was collected from an alluvial aquifer near Globe, Arizona. The aquifer solids consist of unconsolidated alluvium ranging in size from fine sand to coarse gravel, but clay lenses and boulders are also present. Samples were air-dried and sieved with the fraction between 106 and 2000 μm used in the columns. Subsamples were analyzed by X-ray diffraction. The predominant mineral phases, identified in order of intensity, were: quartz > albite > > magnesium orthoferrosilate > muscovite > raguinite > manganese oxide.

The pH_{zpc}, or pH at which the net surface charge of a solid equals zero, is an important parameter affecting both colloid stability and the interaction of mobile particles with immobile matrix surfaces. A mineral's surface charge characteristics, together with surface area, are also important properties affecting contaminant adsorption and coprecipitation. Above the pH_{zpc} minerals possess a net negative surface charge, while below this pH, the net charge is positive. Due to the predominance of silica ($pH_{zpc} \sim 2$) and other minerals such as layer silicates and manganese oxides which have pH_{zpc}'s < 4, most sand and gravel-type aquifer solids exhibit a net negative charge under most environmental pH conditions.

Synthesis and Characterization of Fe_2O_3 Particles. Spherical, monodisperse Fe_2O_3 particles (100-900 nm) were prepared from solutions of $FeCl_3$ and HCl following the method of Matijevic and Scheiner (22). The method was modified by the addition of a spike of $_{26}Fe^{59}Cl_3$, prior to heating, to permit detection of the Fe_2O_3 particles with liquid scintillation counting techniques. This allowed unequivocal discrimination between injected particles and those naturally present and mobilized within the column packing material.

Particle concentration and size ranges utilized in the study were comparable to values reported in the literature for ground waters. Concentrations were determined by both filtration and residue on evaporation techniques. Particle size distributions were determined using photon correlation spectroscopy (PCS) with a 5 mW He-Ne laser as a light source (Malvern AutoSizer IIC). Colloid stability was evaluated using PCS to monitor coagulation. Washed stock colloidal suspensions were used to spike solutions of various electrolyte composition, ionic strength, and pH. Samples were allowed to equilibrate overnight and the pH was readjusted as necessary. An increase in size of the Fe_2O_3 particles indicated instability. In the pH range where the particles were most unstable, coagulation was almost instantaneous. Selected samples were periodically rechecked over several weeks for confirmation of long-term stability. PCS was also used to compare particle size in both influent and effluent column solutions.

Scanning electron microscopy (SEM) was used to confirm both size and shape of the synthesized particles. Figure 1 is a scanning electron micrograph of the ~200 nm synthesized spherical iron oxide particles. The surface area of 200 nm uniformly spherical particles was calculated to be 5.72 m^2/g using the equation,

$$A = \frac{6x10^{-4}}{\rho d} \tag{1}$$

where A is the geometric surface area (m^2/g), ρ the particle density (5.3 g/cm^3), and d the particle diameter (cm).

The surface charge of iron oxides is strongly pH-dependent and the result of acid-base reactions on the particle surface. The pH_{zpc} of the Fe_2O_3 was evaluated from acid-base titrations using varying concentrations of NaCl as the background electrolyte. Electrophoretic mobility (EM) of the colloids was determined using a Rank Brothers Mark II system with a four-electrode capillary cell that was illuminated by a 3 mW Ne-He laser. The system was fitted with a video camera coupled to a rotating prism for the acquisition of mobility data. EM measurements of the particles, in dilute $NaClO_4$, were also used for estimating the surface charge. While more correctly referred to as the isoelectric point (pH_{iep}) of the surface or the pH where the electrokinetic potential of the particle is zero, the two measurements (pH_{zpc} and pH_{iep}) should be approximately equal when performed in the absence of non-specifically adsorbing species (i.e. H^+ and OH^- are the only potential-determining ions in solution). Titrations were performed in a nitrogen filled glove box, whereas the EM measurements were made on the bench in the presence of atmospheric CO_2.

Column Experiments. An Ismatec variable-speed peristaltic pump and Cygnet fraction collector were used with 2.5-cm diameter, adjustable length, glass columns. The columns were carefully packed to bulk densities between 1.5 and 1.7 g/cm^3. Columns were slowly saturated from below and permitted to equilibrate for at least one week prior to a column run. Flow rates were calibrated on an effluent mass to volume basis per unit time. Influent and effluent pH of the suspensions was measured and particle size of both the influent and selected effluent samples was monitored using PCS. Tritiated water was used as a conservative tracer and both it and the iron oxide colloid concentrations were quantified using radioanalytical techniques. Inductively coupled argon plasma and atomic absorption with graphite furnace were used to measure arsenate concentrations.

Results and Discussion

Particle Size Distributions. The primary size distributions of iron oxide particles used in the experiments are illustrated in Figures 2a-b. For the approximately 100 nm particles, the distribution was monomodal and very narrow (102 ± 22

Figure 1. Scanning electron micrograph of synthesized iron oxide particles, ~200 nm diameter.

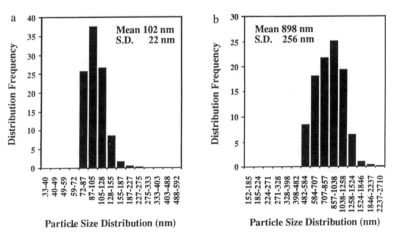

Figure 2. Iron oxide particle size distributions using photon correlation spectroscopy (He-Ne laser light source): (a) ~100 nm; (b) ~900 nm.

nm), and for the larger particles the distribution was also monomodal but somewhat wider (898 ± 256 nm). Colloidal suspensions used in column experiments were always analyzed for size distribution prior to injection and selected effluent solutions were also analyzed for comparison. When the ionic strengths of the suspensions were less than 0.01 M naturally present particles were mobilized within the column. This was reflected by changes in size distribution and increased suspended solids concentrations. It was found that in the higher ionic strength solutions ($\geq$ 0.01 M), particle count, or more correctly photon count, data from the particle size analyzer provided a quantitative measure of particle breakthrough in addition to that provided by scintillation counting of the radio-labeled (Fe^{59}) iron oxide particles.

Zero Point of Charge of Fe_2O_3. The estimated pH_{zpc} from acid-base titrations and electrophoretic mobility measurements was in the pH range of 7.3 to 7.6 (Figures 3a,b). This fundamental double layer property provides a reference point from which to evaluate the particle's charge behavior in natural waters, where changes in aqueous chemistry can alter the effective charge or mobility of the particle. Adsorption of negatively charged organic and inorganic species onto positively-charged iron oxide surfaces has been observed to produce electronegative particles (25-26). This produces an increase in charge repulsion between the particles themselves enhancing colloid stability and favoring particle transport.

Colloid Stability and Mobility. The suspensions were generally stable in up to 0.01 M $NaClO_4$ and NaCl over the pH ranges of 2.0 to 6.5 and 7.6 to 11.0, respectively. The region of instability corresponded to the estimated pH_{zpc}. When suspended in 0.01 M sodium arsenate, sodium phosphate, and sodium dodecyl sulfate, the suspensions were stable throughout the pH range 4 to 10. Specific adsorption of arsenate and phosphate anions onto the initially positively-charged iron oxide particles results in the following:

1) charge reversal at lower pH values,
2) increased net negative charge at some pH's compared to that observed with non-specifically adsorbing ions,
3) increased charge repulsion between particles in suspension, and
4) extension of the pH range over which the particles exhibit a negative mobility.

These phenomena were quantified by means of EM measurements and are illustrated in Figure 4. The net effects are increased stability of particles in suspension and increased electrostatic repulsion with respect to negatively-charged immobile matrix surfaces. Both would contribute to increased particle transport potential in subsurface systems.

In sodium dodecyl sulfate (SDS), as with sodium arsenate and sodium phosphate, the suspensions were stable throughout the pH range 4 to 10; however, the EM values were constant or independent of pH (Figure 5). The

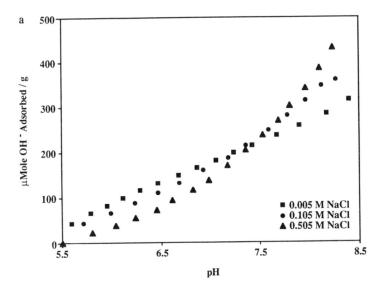

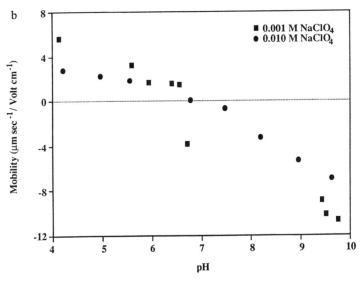

Figure 3. a, Acid-base titrations of Fe$_2$O$_3$ suspensions; b, electrophoretic mobility of Fe$_2$O$_3$ suspensions as a function of pH. (Reproduced from ref. 24. Copyright 1992 American Chemical Society.)

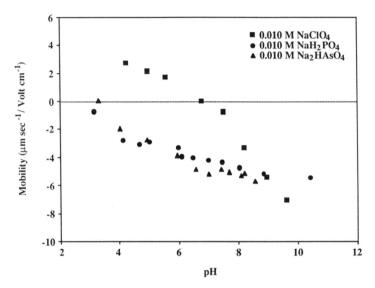

Figure 4. Electrophoretic mobility data in the presence of specific and nonspecific sorbing ions. (Reproduced from ref. 24. Copyright 1992 American Chemical Society.)

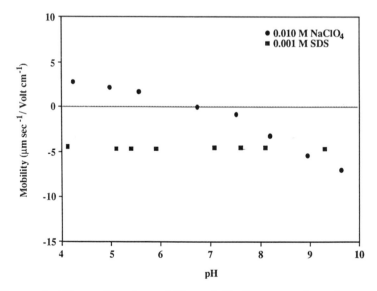

Figure 5. Electrophoretic mobility of Fe_2O_3 suspensions in sodium perchlorate and sodium dodecyl sulfate.

calculated critical micelle concentration (CMC) of the SDS used in the experiments was 0.006 M as determined from turbidity and surface tension measurements. This was very similar to literature-reported values (27,28). The mobility measurements were therefore reflective of the effect of the adsorption of monomers rather than both monomers and micelles. Electrophoretic mobility behavior was analogous to that of a pH-independent charged surface, implying effective occlusion or encapsulation of the normally pH-dependent iron oxide surface. A stearic stabilization mechanism is proposed in contrast to the specific chemical interations involving the phosphate and arsenate anions with the hematite surface. This is under further investigation.

The colloids were generally unstable in sodium sulfate. At low pH the particles readily coagulated as expected due to interaction of the divalent sulfate ion with the positively charged surface sites of the iron oxide. At pH values greater than the pH_{zpc}, the suspensions were only quasi-stable over the pH range 7.6-9.0 (Figure 6). In this range, coagulation proceeded slowly (several hours) affecting apparent EM measurements and thus particle size distributions required careful monitoring during column transport experiments. Several studies have demonstrated that sulfate can broaden the pH range of particle instability (29-31). Hohl et al (32) and Letterman and Vanderbrook (33) have proposed a surface complexation reaction involving hydroxylated surface sites whereby the number of neutral sites is increased or the overall charge is reduced, thus dampening out the crossover from positive to negatively-charged particle surfaces and increasing the region of instability.

Fe_2O_3 Transport. Flow rates used in column particle-transport experiments were comparable to ground water velocities estimated in the alluvium (0.8-3.4 m/d) at the Globe, Arizona site. The injected Fe_2O_3 particles generally broke through at the same time or prior to the tritium (Figure 7). The rate of Fe_2O_3 (with adsorbed arsenate) transport through the packed columns was over 21 times faster than for the dissolved arsenate (Figure 8). A summary of column results for the particle transport experiments are compiled in Table I. No Fe_2O_3 transport occurred on the positive side (lower pH) of the pH_{zpc} indicating an attractive interaction with the column matrix material (see run 1). In low ionic strength suspensions of NaCl and $NaClO_4$, transport exceeded 50% of initial particle concentrations (runs 2 and 3). These suspensions could generally be considered more representative of natural ground waters than those containing only phosphate or arsenate anions. There was substantially lower particle breakthrough using the sulfate-based suspensions presumably due to the relative instability of the suspensions (runs 4 and 5).

Maximum particle breakthrough occurred with the SDS, phosphate and arsenate-based suspensions and appeared to be independent of the flow velocities, particle concentrations or column lengths used in the various column experiments (runs 6 through 12). The specific adsorption of the predominantly divalent phosphate and arsenate anions onto the Fe_2O_3 surface produced an increase in the charge repulsion between individual particles in suspension. The effect is a lowering of the pH_{iep} and production of a wider pH range where

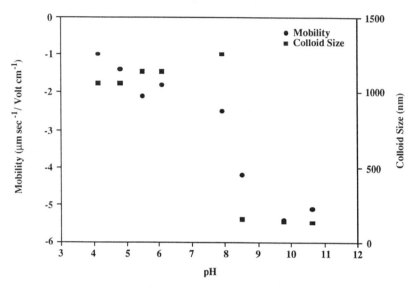

Figure 6. Colloid mobility and stability in 0.01 M sodium sulfate as a function of pH.

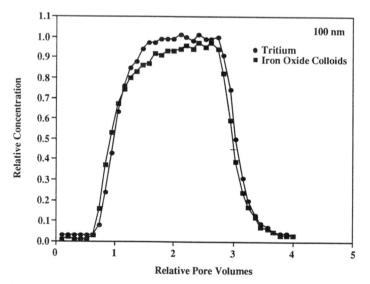

Figure 7. Iron oxide particle breakthrough in 0.01 M sodium arsenate suspensions compared to tritium breakthrough (pH 7.6).

Table I. Results of colloidal Fe_2O_3 transport through natural aquifer material

Run	Size (nm)	pH	Velocity (m/d)	Part.Conc. (mg/L)	Ionic Strength	Anion	%C$_o$ thru	Col.L. (cm)
1	200	3.9	3.4	10	0.005	Cl^-	0	3.8
2	125	8.9	3.4	10	0.005	Cl^-	54	5.1
3	150	8.1	3.4	5	0.001	ClO_4^-	57	3.8
4	250	8.1	3.4	10	0.03	SO_4^{2-}	17	3.8
5	150	8.9	3.4	5	0.03	SO_4^{2-}	14	3.8
6	100	7.6	3.4	5	0.03	$HAsO_4^{2-}$	97	2.5
7	100	7.6	1.7	5	0.03	$HAsO_4^{2-}$	96	2.5
8	100	7.6	0.8	5	0.03	$HAsO_4^{2-}$	93	2.5
9	125	7.6	3.4	10	0.03	HPO_4^{2-}	99	5.1
10	100	7.6	1.7	5	0.03	HPO_4^{2-}	99	2.5
11	100	7.6	3.4	5	0.03	HPO_4^{2-}	99	2.5
12	100	7.6	3.4	5	0.01	SDS	98	2.5
13	900	7.0	3.4	50	0.03	HPO_4^{2-}	33	3.8
14	900	7.0	3.4	50	0.03	HPO_4^{2-}	30	3.8

the particles are negatively charged. As a result, the colloids are more stable in suspension and, as electronegative particles, have less interaction with negatively-charged immobile matrix surfaces. Although the transport behavior in SDS was similar to that in arsenate and phosphate suspensions, the surface interaction which produced stable and constant negatively-charged particles over a wide pH range is probably different and under further investigation.

Particle size had an inverse effect on breakthrough; that is, there was increasing Fe_2O_3 breakthrough with decreasing size. While the larger particles were still transported (runs 13 and 14), there were significant differences between the 100 nm and 900 nm size classes. A complicating factor in resolving these differences was the use of a different aquifer sample (well 452) for the larger colloids (runs 13 and 14). Particle size distributions of the two aquifer samples were significantly different (Figure 9); however, XRD analyses showed no significant differences in mineralogy. Thus, the surface chemistry should not be altered.

All of the above parameters (Table I. headings) or variables were explored in detail with the SAS program JMP, using %C$_o$ breakthrough as the response variable and data from twenty-one column runs. The 900 nm particles were not part of that analysis due to the different particle size distributions of the two samples. Only colloid size and anion significantly influence the %C$_o$ breakthrough. Combining these two parameters into a two-way main effects analysis of variance (2-way ANOVA) model accounts for 98.4% of the variability in the colloid breakthrough results. All other factors tested provided no significant correlation over the parameter ranges utilized in this study.

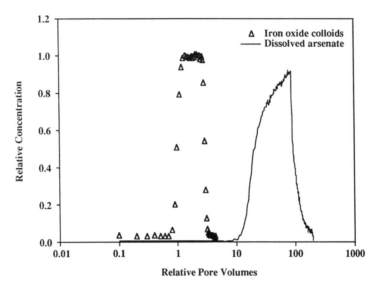

Figure 8. Comparison of colloid associated arsenate and dissolved arsenate breakthrough (pH 7.6).

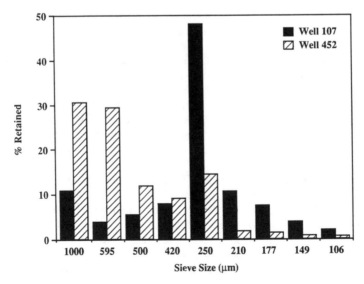

Figure 9. Size fractionation comparison for Globe, AZ aquifer solids.

Summary and Conclusions

Laboratory experiments using natural aquifer material and representative inorganic colloids indicate that the transport of colloidal material through sand and gravel-type aquifers may be significant under certain hydrogeochemical conditions. The production of strong surface charge repulsive forces or steric effects due to the adsorption of inorganic or organic anions on iron oxide surfaces have been shown to stabilize colloids and promote particle transport in porous media. Due to the strong reactivity of many inorganic colloids in natural subsurface systems the potential for these colloids to transport adsorbed contaminants may be significant.

Additional work is needed using more realistic electrolyte mixtures and concentrations for proper simulation of typical dilute ground waters. While such waters can be expected to produce much less dramatic particle transport results, this transport mechanism may still be significant, particularly for highly toxic or radioactive species.

Acknowledgements

The authors wish to acknowledge and thank Cynthia J. Paul for her considerable technical support, Donald A. Clark for radioanalytical and inorganic analyses, and Terry F. Rees for scanning electron microscopy analyses and helpful guidance and comments.

Disclaimer

Although the research described in this article has been funded wholly or in part by the United States Environmental Protection Agency, it has not been subjected to the Agency's peer and administrative review and therefore may not necessarily reflect the views of the Agency and no official endorsement should be inferred. Mention of trade names or commercial products does not constitute endorsement or recommendation for use.

Literature Cited

(1)   Ross, S.; Morrison, I.; *Colloidal Systems and Interfaces*. John Wiley & Sons,1988.

(2)   Buddemeier, R.W.; Hunt, J.R. *Applied Geochemistry* 1988, *3*, 535-548.

(3)   Ryan, J.N.; Gschwend P.M. *Water Resour. Res.* 1990, *26*(2), 307-322.

(4)   Puls, R.W.; Eychaner, J.H. In *Fourth National Outdoor Action Conference on Aquifer Restoration, Ground Water Monitoring and Geophysical Methods*, National Water Well Association, Dublin, OH, 1990.

(5)   Sheppard, J.C.; Campbell, M.J.; Kittrick, J.A. *Environ. Sci. Technol.* 1979, *13*(6), 680-684.

(6)   Takayanagi,K.; Wong, G.T.F. *Marine Chem.* 1984, 14, 141-148.

(7)   Sandhu, S.S.; Mills, G.L. *Kinetics and Mechanisms of the Release of Trace*

Inorganic Contaminants to Ground Water from Coal Ash Basins on the Savannah River Plant, Savannah River Ecology Lab, Aiken, SC, DOE/SR/15170-1, 1987

(8) Means, J.C.; Wijayaratne, R. *Science* 1982, *215*(19), 968-970.
(9) Nelson, D.M.; Penrose, W.R.; Karttunen, J.O.; Mehlhaff, P. *Environ. Sci. Technol.* 1985, *19*, 127-131.
(10) Buddemeier, R.W.; Rego, J.H. *Colloidal Radionuclides in Groundwater. FY85 Annual Report.* Lawrence Livermore National Laboratory, Livermore, CA, UCAR 10062/85-1, 1986.
(11) Gschwend P.M. and M.D. Reynolds, *J. of Contaminant Hydrol.* 1987, *1*, 309-327.
(12) Nightingale, H.I.; Bianchi, W.C. *Ground Water* 1977, *15*(2), 146-152.
(13) Eichholz, G.G.; Wahlig, B.G.; Powell, G.F.; Craft, T.F. *Nuclear Tech.* 1982, *58*, 511-519.
(14) Champlin, J.B.F.; Eichholz, G.G. *Health Physics*, 1976, *30*, 215-219.
(15) Tillekeratne, S., Miwa, T.; Mizuike, A. *Mikrochimica Acta* 1986, *B*, 289-296.
(16) Champ, D.R., Merritt, W.F.; Young, J.L. *Potential for Rapid Transport of Pu in Groundwater as Demonstrated by Core Column Studies.* In Scientific Basis for Radioactive Waste Management. Vol. 5, Elsevier Sci.Publ., NY, 1982.
(17) Reynolds, M.D. *Colloids in Groundwater.* Masters Thesis. Mass. Inst. of Tech. Cambridge, MA, 1985.
(18) Harvey, R.W.; George, L.H.; Smith, R.L.; LeBlanc, D.R. *Environ. Sci. Technol.* 1989, *23*(1),51-56.
(19) Enfield, C.G.; Bengtsson, G. *Ground Water* 1988, *26*(1),64-70.
(20) Matijivec, E.; Kuo, R.J.; Kolny, H. *J. Colloid Interface Sci.* 1980, *80*(1), 94-106.
(21) Champlin, J.B.F.; Eichholz, G.G. *Health Physics* 1976, *30*, 215-219.
(22) Matijevic, E.; Scheiner, P. *J. Colloid Interface Sci.*, 1978, *63*(3), 509-524.
(23) Tipping, E.; Cooke, D. *Geochim. Cosmochim. Acta* 1982, *46*, 75-80.
(24) Puls, R.W.; Powell, R. *Environ. Sci. Technol.* 1992, in press.
(25) Boyle, E.A.; Edmond, J.M.; Sholkovitz, E.R. *Geochem. Cosmochim. Acta* 1977, *41*, 1313-1324.
(26) Liang, L.; Morgan, J.J. *Aquatic Sciences*, 1990, *52*(1), 32-55.
(27) Preston, W.C. *J. Phys. Colloid Chem.* 1948, *52*, 84-90.
(28) Chiu, Y.C.; Wang, S.J. *Colloids and Surfaces* 1990, *48*, 297-309.
(29) Hanna, G.P.; Rubin, A.J. *J. Am. Water Wks. Assoc.* 1970, *62*, 315-321.
(30) Snodgrass, W.J.; Clark, M.M.; O'Melia, C.R. *Water Res.* 1984, *4*, 479-488.
(31) Packham, R.F. *J. Colloid Sci.* 1965, *20*(1), 81-92.
(32) Hohl, H.; Sigg, L.; Stuum, W. *Symposium on Particulates in Water*, 175th ACS National Meeting, Anaheim, CA. 1978.
(33) Letterman, R.D.; Vanderbrook, S.G. *Water Res.* 1983, *17*, 195-204.

RECEIVED December 18, 1991

Chapter 5

Colloid Transport and the Gas–Water Interface in Porous Media

Jiamin Wan and John L. Wilson

Department of Geoscience and Geophysical Research Center, New Mexico Institute of Mining and Technology, Socorro, NM 87801

The influence of the gas-water interface on the transport of colloidal sized particles through porous media was experimentally studied in micromodels and columns. The interfaces were created by trapping gas bubbles in the pore space. Several variables were tested: particle hydrophobicity, particle charge, and ionic strength of the solution. The gas-water interface adsorbed negatively and positively charged polystyrene latex particles and clean clay particles, over a wide range of ionic strengths. The degree of this adsorption increased with ionic strength. Positively charged particles had more affinity for the interface. Once particles were adsorbed onto the gas bubbles, it was difficult to detach them by shear stress. This adsorption was also irreversible to changes in ionic strength. A gas-water interface in motion effectively stripped particles from solid surfaces and carried them along. These results suggest that gas water interfaces play an important role in colloid transport, that might be manipulated to enhance aquifer remediation.

Gas-aqueous interfaces are common in subsurface environments. In the vadose zone a continuous gas phase shares pore space with the aqueous phase. In the saturated-zone gas bubbles may be generated by various processes: entrapment of air as the water table fluctuates, organic and biogenic activities, or gas coming out of the solution as the aqueous phase pressure drops.

Colloid transport is concerned with the movement and attachment/detachment of colloidal sized particles (≤ 10 μm) in porous aquifer materials. Interest in this subject has grown because of the possibility that organic and inorganic pollutants

0097–6156/92/0491–0055$06.00/0

may adsorb onto mobile colloids, dramatically enhancing pollutant mobility. Work on colloid transport has focused on porous materials saturated with a liquid. In this paper we examine the role of the gas phase and, in particular, the gas-aqueous interface in colloid transport. We also consider the application of the special properties of the interface as a potential component of aquifer remediation schemes.

Adsorption of large-size mineral particles (≥ 100 μm) on gas bubbles has been intensively studied in the field of flotation (1-4) because of its economic importance. Flotation involves the attachment of so-called hydrophobic particles to air bubbles and the subsequent transfer of the particle-laden bubbles to the froth. The froth contains a concentration of mineral particles that is skimmed off. The probability of flotation can be described in the general form (3):

$$P = P_c P_a P_s \qquad (1)$$

P_c is the probability of collision between a particle and a gas bubble. P_a represents the probability of adhesion of the particle to the bubble in the time of contact. During that short time, the liquid disjoining film must elongate, then rupture and recede. P_s is the probability of formation of a stable bubble-particle aggregate, and is a function of the contact angle. The greater the contact angle, the stronger the particle-bubble aggregate. In flotation theory, "hydrophobic" refers to all non-zero contact angles.

The general thermodynamic criterion (Eqn. 1) of flotation can be borrowed to describe the processes in the present research, although there are important differences. First, the mineral particles of concern in flotation are much larger than colloids. Second, the surface of these mineral particles is usually covered with adsorbed surface-active agents, called collectors, that impart a hydrophobic character to the particles. The surface of the bubbles is stabilized by surface active frothing agents. The attachment of particles and bubbles is proceeded by interactions between molecules of these two reagents. Natural porous media systems are essentially free of surface active agents. Third, flotation applications occur in a reactor vessel, whereas this study concerns porous media.

Methodology

Micromodel Experiments. An optical microscope and a glass micromodel (an artificial pore network) were used to directly observe fluid and particle transport, and the interactions among particles and interfaces, on a pore scale. The glass micromodels were created by etching a pattern onto two glass plates which were then fused together in a furnace (5). Although the pore network was two dimensional, the pores had a complex three-dimensional structure (6). Two of the micromodel patterns used in the experiments are shown in Figures 1 and 2. The pore networks were composed of pore bodies, connected together by pore throats. Pore wedges composed the corners of pore bodies and throats, and are a feature of the third dimension that is not readily apparent in the figures. The width of pore throats in these models was 100 to 160 μm. The diameter of the pore bodies was about 320 μm. Their depth varied from 50 μm to 200 μm. The total pore volume of the different micromodels ranged from 0.14 to 0.17 ml. Each micromodel was mounted horizontally on the stage of a microscope (Zeiss Axiophot). The microscope

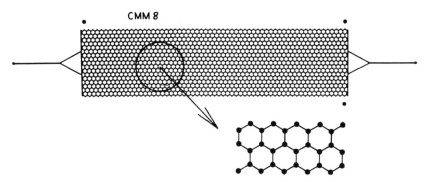

Figure 1. A micromodel pattern

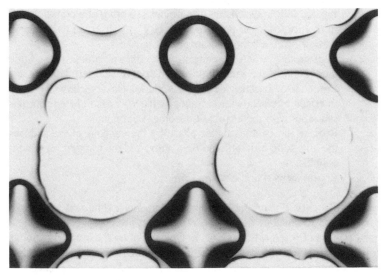

Figure 2. A pore network under a microscope. Every pore body had a trapped air bubble. The pore bodies were about 350 μm in diameter in this network.

had a fluorescence lighting system, dark field image capability, and long working distance objectives. Photomicrographs and video tapes could be taken simultaneously. The flow rate was precisely controlled by a 50 cc syringe pump (Harvard Apparatus, model 4400-001,South Natick,Mass). Both artificial and natural particles were used: surfactant-free polystyrene latex particles (Interfacial Dynamics Corp.,Portland, Oregon) and clay particles (Clay Mineral Society Source Clay Mineral Repository, Columbia, Missouri). The fluorescent sulfated polystyrene particles were negatively charged and 1.05 μm in diameter, while the fluorescent amidine polystyrene particles were positively charged and 0.60 μm in diameter. The montmorillonite and kaolinite particles were sorted with a centrifuge. Particle sizes, measured with a scanning electron microscope, were less than or equal to 0.5 μm. Clay particles were washed carefully before use. All suspensions were monodispersed and examined through the microscope before use. Sodium nitrate was used to adjust the ionic strength, and sodium bicarbonate buffered the pH of the aqueous solutions. The inorganic chemicals were analytical reagent grade. Solutions and suspensions were prepared using Milli-Q water.

Six micromodel experiments were performed. The experiments were designed to study how the types of particles, the signs of the particle charge, and the aqueous chemistry affected interactions among particles and gas-aqueous interfaces. The procedure followed in the micromodel experiments was:

a.   Saturate a clean micromodel with particle free solution, then trap air bubbles as a residual phase wing capillary forces (see Figure 2).

b.   Inject a dilute suspension of particles (10 ppm) at a constant rate (1.5 ml/hr) for 20 to 50 pore volumes (PV's). Observe attachment of the particles to the solid and air bubble surfaces.

c.   Replace the particle suspension with either a particle-free solution or pure water at the same flow rate. Observe selected detachment.

d.   Inject air to replace the aqueous phase. Observe how the air-water interface interacts with particles as the water is reduced to a residual saturation and the air phase becomes continuous.

e.   Replace the continuous air phase by the aqueous phase and observe the migration of bubble-particle units, and the trapping of residual air bubbles.

f.   Repeat steps d and e several times.

Column Experiments.   For the column experiments cylindrical glass columns (ACE glass Inc.), 30 cm long and 5 cm in diameter, were packed with a uniform quartz sand (Unimin Co.,New Canaan,CT). One end of each column was connected to a syringe pump (ISCO Model 500) which controlled flow rates precisely. The other end was connected to a fraction collector. The sand grain sizes ranged in size from 250 to 300 μm in diameter. The sand was cleaned by soaking it for 20 minutes, first in 10% HF and then in 0.1 N NaOH, with a distilled water rinse after each. It was ultrasonicated for 20 minutes in distilled water, and rinsed again. The turbidity and pH of the rinse water were measured. If they were not close to the value for distilled water the procedure was repeated. The sand was dried in an oven at 70°C.

Surfactant-free sulfated polystyrene latex particles, 0.22 μm in diameter, were used in the column experiments. Two column experiments were always carried out in parallel. The experimental conditions were the same in both columns, except that one was completely saturated with water, and the other had trapped gas. The experimental procedure was:

a.   Pack the two columns at the same time under the same conditions.

b.   Dissolve any air trapped during packing with de-gassed water to obtain completely water saturated columns, then calculate the bulk densities, pore volumes, and porosities of both columns.

c.   Flood the columns with particle-free solution (0.001 M N_aNO_3) to obtain the desired chemical conditions for both columns.

d.   For one of the two columns, drain the aqueous phase with nitrogen gas, then imbibe with particle-free solution, to entrap nitrogen gas as bubbles.

e.   Inject three pore volumes of the dilute particle suspension (10 ppm) with dissolved bromide tracer at a mean velocity of ten centimeters per hour for a few pore volumes. Bromide was used as a conservative tracer.

f.   Replace the particle suspension with particle-free solution or pure water at the same velocity.

The pH values of all solutions were controlled at 7.0 and varied from   6.9 to 7.1. Bromine was analyzed with a bromide electrode (Orion Research Inc.). The concentrations of particles were measured with a turbidimeter (Monitek Nephlometer model 21). The sensitivity of the turbidimeter was 0.01 ppm for the polystyrene particles with a diameter of 0.22 μm.   The injected solution was saturated with nitrogen gas. The gas bubble volume saturation in the column was monitored and remained constant during the test.

Results and Discussion

Micromodel Experiments.   The first three micromodel trials tested the effects of ionic strength using fluorescent sulfated polystyrene particles (1.05 μm). In trial 1, the ionic strength was 0.001 N. The particles were preferentially adsorbed on the air bubbles, with very few adsorbing on the glass surface, as shown in Figure 3. The photomicrograph was taken after flooding with 30 pore volumes of the suspension, then replacing it with a particle free solution. This behavior indicates that the net energy is repulsive for the glass/particle pair, but attractive for the bubble/particle pair. If an air bubble has a negative surface potential (7-11), and the Hamaker constant of the system is positive (12; see below), then both van der Waals and electrostatic interaction should be repulsive. There must be either a third, so-called structural force (13-22), which is larger than the positive van der Waals and electrostatic forces, or sufficient particle kinetic energy, to overcome this barrier. Both of these hypotheses are discussed later.

In trial 2, the ionic strength was increased by a factor of 100, while the other conditions remained the same. More particles were adsorbed on both the glass surface and the air-bubbles than in trial 1. Figure 4 is a photomicrograph taken after the injection of 30 pore volumes of particle suspension and replacement of the suspension with a particle free solution. Compared with trial 1, the total potential of the glass/particle pair has changed sign, due to compression of the double layer. The bubble/particle pair experienced an increased attractive force, indicating that the electrostatic component also plays a role at the air-water interface. The particle-free solution flow rate was then increased by a factor of 10. Some of the particles desorbed from glass surface because of increased shear forces. However, there was no desorption from the air bubbles, which captured even more particles from solution, as shown Figure 5. It is clear that detaching particles from the air bubbles by flow is more difficult than from glass surface. The next step involved injecting air into the micromodel to displace some of the liquid and the originally trapped air bubbles. The air-water interface migrating into the micromodel efficiently stripped particles from the glass surface. Particle-free solution was then injected to remove most of the air. The moving air-aqueous interface detached and carried away many of the particles. Detachment from the glass surface was most complete in pore bodies and large throats, and was not very effective in deep pore wedges, fine throats, and dead zones (areas where little flow occurs due to pore structure), since air was not able to reach there. Figure 6 illustrates the results after three cycles of air/water replacement. The glass surface is clean. Most of the attached particles have disappeared. A small residual air bubble in a pore body is covered by particles, implying increased adsorption caused by a large probability of collision, P_c.

The experimental conditions in trial 3 were the same as those in trial 2, except pure water was used, instead of particle-free solution, to replace the suspension. After running 30 pore volumes of pure water through the micromodel at the same flow rate, some of the particles attached on the glass surface were released. No significant desorption from the air bubbles was observed. Adsorption of particles on the glass was partially reversible, while adsorption on the air bubbles was irreversible, by changing ionic strength. The energy valley of bubble and particle pair is deeper than that of glass and particle pair, perhaps because of the structural force, or because of capillary forces.

In trial 4, positively charged amidine polystyrene particles (0.60 μm) were suspended in a low ionic strength solution (0.001 N). The other experimental conditions were the same as those in trial 1. As expected, the particles were strongly adsorbed on the glass surface. The air bubbles also preferred the positively charged particles over the negatively charged ones. The two sides of both pairs in this system had opposite charge signs. This changes the energy barrier to a valley for the glass/particle pair and, assuming a third force, increases the depth of the energy valley for the bubble/particle pair. After air was injected, some of the particles were stripped from the glass walls, as shown in Figure 7. Kaolinite and montmorillonite particles were used in trials 5 and 6, respectively. The ionic strength of the solutions was adjusted to 0.001 N and the pH was controlled at 7.0. Clay particles were monodispersed and checked with a dark field microscope.

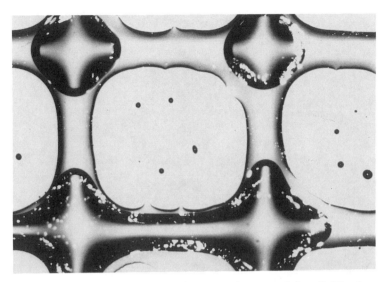

Figure 3.   Fluorescent sulfated polystyrene latex particles (1.05um) were preferentially adsorbed on the air bubbles rather than the glass surface in the 0.001M $NaNO_3$ solution. The photomicrograph was taken after the passage of 30 pore volumes of dilute suspension and it s replacement by a particle free solution. The air bubbles were about 200 μm in diameter.

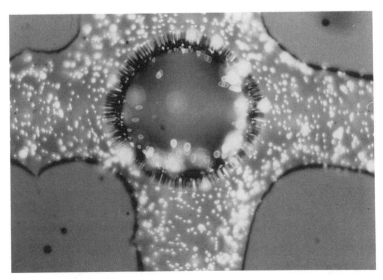

Figure 4.   Fluorescent sulfated polystyrene latex particles (1.05um) were adsorbed on both the air bubble and the glass surface in a 0.10 M $NaNO_3$ solution. The photomicrograph was taken after a passage of 30 PV of dilute suspension and it s replacement by particle free solution. The diameter of the air bubble was about 200 μm.

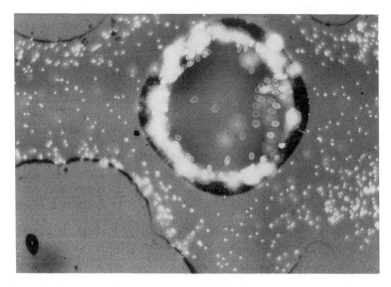

Figure 5. Same experiment as Fig 4, after the passage of 30 PV of particle free solution at a flow rate 10 times higher. Some of the particles were detached from glass surface by shear stress, however, it appears that some of these detached particles accumulated on the air bubbles.

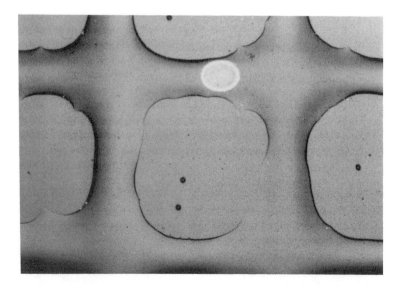

Figure 6. Same experiment as Figs 4 and 5, after three cycles of air phase and water phase replacement. Almost all of the particles adsorbed on glass surface were removed by moving air-water interfaces. A small residual air bubble was wholly covered by particles.

These experiments were difficult to observe because the monodispersed particles were very small and were without a fluorescent dye. The dark field technique was less than ideal because the three phases in the micromodel, glass, water and air, reflected light differently. Nevertheless, it was observed that the clay particles preferentially accumulated on the air bubbles even in the low ionic strength solution. Figures 8 and 9 were taken under transmitted light conditions. Both photomicrographs show montmorillonite particles accumulating on an air bubble. Figure 8 shows the down-stream part of an air bubble. Figure 9 shows the greater curvature portion of a long air bubble in a pore throat. Particles preferred these low flow rate areas. Kaolinite particles behaved similarly.

Rapid oscillations of some particles on the bubble surface were observed in all of the micromodel experiments. This was apparently related to both Brownian motion and shear stress in the flowing aqueous phase. Particles also tended to accumulate on the downstream portions of the bubbles at higher flow rates.

Column Experiments. The column experiments were designed to complement the micromodel experiments. For the suspension, 10 ppm of sulfated polystyrene latex particles (0.22 μm) were monodispersed in a solution of 0.001 M $NaNO_3$. After three pore volumes of the suspension, pure water was injected at the same velocity to replace the suspension. Figure 10 shows the breakthrough curves (BTC's) from the column saturated with water. The classical step-function BTC for bromine implies that the experiments were well controlled. There was no particle breakthrough until the front of pure water reached the column exit. At that time, a sharp particle breakthrough appeared with a peak concentration about five times that of the influent concentration. As the ionic strength was very high, it was believed that all the particles were adsorbed on the surface of quartz grains. When the ionic strength was reduced by the pure water, some of the particles were released from the quartz surface instantly as a result of double layer expansion.

Figure 11 shows BTC's from the column with trapped gas bubbles. Again the BTC for bromine followed the classical response to a step-function input. The particle breakthrough appeared at the same time that the front of pure water reached the column exit. However, the peak particle concentration in this case was about 6 times lower than before. Fewer particles were released by the reduction of ionic strength. This implies that the gas bubbles trapped a large proportion of the particles, and the gas-water interface did not respond to the reduction of ionic strength. The small amount of particles released came from the solid surface. The results are consistent with the micromodel experiments.

Discussion. DLVO theory and the extended DLVO theory (15-18,21-23) provide a theoretical foundation for discussing the results. DLVO theory is based on the assumption that the forces between two bodies in a liquid can be regarded as the sum of two contributions: the Van der Waals force and the electrostatic force. This theory is an example of a continuum theory --- both the interacting bodies and medium are characterized by their dielectric properties and surface effects are not taken into account. The theory breaks down for very hydrophobic substances, as clearly stated by the original authors (16). Many investigators have provided both theoretical and experimental evidence for the existence of forces other than the classic DLVO forces (15-23). Churaec and Derjaguin (18) referred to these so-

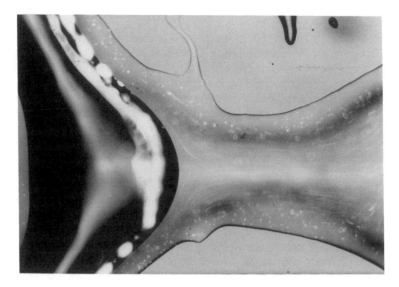

Figure 7. Positively charged amidine polystyrene latex particles (0.6um) being stripped from the glass surface by air-water interfaces. This photomicrograph was taken after the injection of air to replace the aqueous phase. The pore body was about 400um in diameter.

Figure 8. Montmorillonite particles accumulated on the down-stream side of an air bubble. The air bubble was about 300um in diameter, and the sizes of particles were $\leq$ 0.5 μm. The photomicrograph was taken under transmitted light.

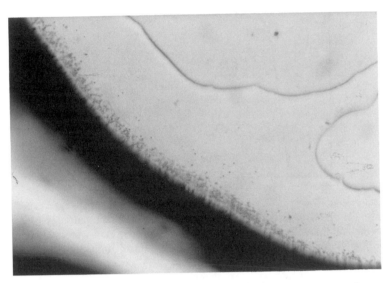

Figure 9. Montmorillonite particles accumulated on an air bubble in a pore throat.

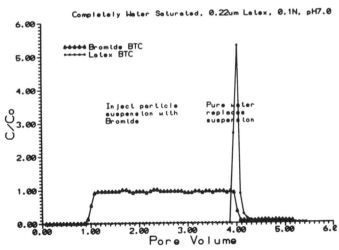

Figure 10. BTC's of ideal tracer and sulfated polystyrene particles (0.22 µm) for the column completely saturated with water. Pure water was introduced after 3 PV's of particle suspension. The suspension had 10 ppm of latex particles and 0.1 M NaNO₃.

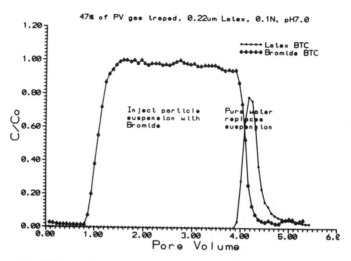

Figure 11. BTC's of ideal tracer and sulfated polystyrene particles (0.22um) from the column with 47% of PV trapped with N_2 gas. Pure water was injected after 3 PV's of particle suspension. The suspension had 10ppm of latex particles and 0.1 M $NaNO_3$.

called non-DLVO forces as structural forces, and suggested they can be repulsive between two interacting hydrophilic surfaces and attractive for hydrophobic surfaces. The origin of structural forces is believed to be the overlapping of the boundary layers of surfaces when approaching one another. However, many related problems remain unsolved: e.g. what is the special structure within the boundary layer? How do the boundary layers interact with each other? Some investigators (19,20,21,24,25,26) suggest that the non-DLVO forces arise mainly from the configurational rearrangement of water molecules in the vicinity of the surfaces, while others believe they are due to phase changes within the interlayer confined by the two approaching surfaces (17,27,28) or anomalous polarization of water molecules in the boundary layer (19,29,30). Derjaguin and Zorin (31) were the first to obtain isotherms for structural forces. The isotherm represents the dependence of the structural forces on the thickness, h, of a liquid interlayer between the surfaces confining it (1,17) and is described by:

$$V_s(h) = K \exp(-h/l) \tag{2}$$

where both K and l are constants. The values of K are positive for hydrophilic systems, where the structural forces cause repulsion of the surfaces. The range of the repulsive force decreases as the wetting conditions grow more hydrophobic. The first theory of the structural forces, developed by Marcelja and Radic (15), as well as its subsequent modifications (1,17,21,22), relate the value of the constant K to the ordering state of the dipole molecules of the liquid. The values of "l" are related to the corresponding correlation length and they decrease as the values of the contact angle increase. Therefore, the extended DLVO theory contains three components:

Van der Waals, electrostatic, and structural forces. The interplay of these three forces determines the height of the potential barrier (if any) preventing adsorption of the colloids onto the interface. Any particle surmounting the potential barrier is drawn into a very deep capillary well and remains irreversibly adsorbed (12).

Referring to the experiments herein, there are several points worth discussing. The first point is that the adsorption of particles onto the glass surface can be explained by classical DLVO theory. This point is taken from the column experiments and the comparisons of the micromodel results of trials 1 to 2, and 1 to 4, which show that both ionic strength and the sign of charges carried by the particles strongly affect the adsorption of particles on the solid surface. A second point is that when gas-water interfaces are involved, even extended DLVO theory cannot adequately explain the phenomena. As we have seen in trials 5 and 6, both kaolinite and montmorillonite are negatively charged, hydrophilic particles, and the air-water interface is also slightly negatively charged (7-11); hence, the electrostatic and structural terms in the generalized DLVO theory both result in repulsive forces. For the Van der Waals force in this case, the Hamaker constant is

$$A(132) = (\sqrt{A11} - \sqrt{A33})(\sqrt{A22} - \sqrt{A33}) \tag{3}$$

Since water exhibits a Hamaker constant, $A(33)$, intermediate in value to that of air, $A(22)$, and most solids, $A(11)$, constant $A(132)$ must be negative, i.e., the Van der Waals force is also repulsive (12).

Although there is no attractive force predicted by extended DLVO theory, there is still strong adsorption of hydrophilic clay particles onto the gas-water interface. One possibility is that the existing expression for the structural force is not quite complete. Another possibility is that there exists a so-called "channel effect", in that the particles with sufficient kinetic energy will penetrate the energy barrier and reach the air-water interface, where they will be trapped by capillary forces. The probability of this penetration depends on the particle kinetic energy and the height of the barrier. Once a particle penetrates the barrier it will be retained by capillary forces if it has any finite contact angle (12, 32). This is certainly the case for the polystyrene particles, but what is the contact angle for the clean clays?

Implications

The experiments described above focused on the idealized geometry of isolated gas bubbles trapped by capillary forces in an otherwise water filled pore space. Particles attached to these gas-water interfaces are no longer able to move. In the vadose zone, the gas phase and the gas-water interface, are both interconnected. Particles attached to the interface may be highly mobile. If we assume a fluid mechanics model with zero shear stress at the gas-water interface, then water velocities near the interface will be greater than the average water velocity. Depending on conditions, particles on the interface may then move faster than particles suspended in the water nearer the pore walls, were velocities are lower. Thus the attraction of particles to the interface, coupled with the higher velocity at the interface, could dramatically enhance particle mobility.   Further experiments are required to test this hypothesis. During unsteady flow, when fluid saturations are changing, there are additional considerations. For example, as water infiltrates into a dry soil, moving interfaces may collect particles from the pore wall and mobilize them, carrying them deeper into the aquifer. This process would enhance the movement of organic colloidal matter from the soil's active-A-horizon deep into the vadose zone. Evapotranspiration of soil water, drying the soil, would then re-deposit the particles on the pore walls. In another example, a fluctuating water table is expected to carry mobilized particles with it. Sweeping up and down over a portion of the aquifer, it would alter the distribution and concentration of particles, relative to that found deeper in the aquifer, or in the vadose zone above. One speculation is that particle concentrations will be depleted in this interval, because of the repeated interfacial intrusions. An opposing speculation is that it simply enhances the downward movement of new particles arriving from above, at least in groundwater recharge areas. In groundwater discharge areas, with water moving up through the vadose zone, it suggests that colloid concentrations should be small except following an unusually low water table fluctuation.

In contaminated aquifers there may be non-aqueous phase liquids present that provide liquid-liquid interfaces (5,6). Once the main body of non-aqueous phase liquid passes it leaves behind a trail of capillary trapped blobs, analogous to the gas bubbles in these experiments. Although the interfacial force balance will be different, these results suggest that colloids will be attracted to the interface. This includes clays and oxides that normally coat the pore walls. The mobilization of these materials could alter interfacial properties and their redeposition can restruct the pore space. Both mechanisms would lead to a change in behavior of the non-aqueous liquid, and the efficacy of various aquifer remediation techniques. Bacteria constitute one colloid of concern in remediation. It appears that bacteria will be perferentially attached to the liquid-liquid interface.

It may be possible to use the attraction of colloids to fluid-fluid interfaces in order to assist with aquifer remediation or to prevent future contaminant migration. Consider multiple wetting-drying cycles, such as employed in the micromodel experiments. These could be used to deplete an aquifer zone of contaminated colloidal material, or perhaps to remove colloidal material prior to its exposure to contaminated water. In another example, suppose it is desired to transport bacteria or other micro-organisms to a zone contaminated with trapped non-aqueous phase liquid. Water chemistry can be controlled to maximize their mobility, yet they will be adsorbed by the liquid-liquid interfaces in the contaminated zone. Then the water chemistry can be adjusted to maximize population growth and degradation of the non-aqueous phase liquid. In a similar way colloid capsules could be used to deliver chemical packages to the blobs, perhaps to enhance their solubility.

Conclusions

The contribution of this work to the understanding of particle transport during gas and water fluid flow in the porous media may be summarized as follows:

1. Gas-water interfaces always adsorb particles: negatively and positively

charged polystyrene latex particles and clean clay particles, at high and low ionic strength conditions (0.1 to 0.001 M NaNO$_3$ in this study). The degree of this adsorption increases with ionic strength. Positively charged particles have more affinity for the gas-water interface.

2. The adsorption of particles on gas bubbles is irreversible with a change of ionic strength. It is also difficult to detach the adsorbed particles from gas bubbles by shear stress. Increasing flow velocity causes some particles randomly distributed on interface to concentrate on the downstream side of a bubble and in areas of greater curvature, and also causes some particles to oscillate rapidly along the gas-water interface. Once adsorbed onto the interface few particles can be removed by flow unless the interface itself is moved.

3. The observed interactions between glass surfaces and particles are conceptually consistent with the classic DLVO theory. The large attractive force between gas bubbles and particles in water can be explained by a hydrophobic structural force, although a better interpretation for the adsorption of hydrophilic particles is necessary. Kinetic energy considerations can also explain these observations, if one assumes that even the clean clays have a finite contact angle.

4. Moving gas-water interfaces can strip and carry away particles.

Acknowledgments. The authors acknowledge the Subsurface Science Program of the U.S.Department of Energy (Contract G04-89ER 60829) for supporting this work.

Literature Cited

1. Derjaguin, B.V.; Dukhin, S.S. In *Mineral Processing: Developments in Mineral Processing*; Laskowski, J.,Ed.; Elsevier: Amsterdam, Polish Scientific Publishers, Warsaw, 1981, Vol.2, p.21.
2. Jameson, G.J. In *Principles of Mineral Flotation*; Jones,M.H.; Woodcock, J.T., Ed.; The Australia Institute of Mining and Metallurgy, Victoria, Australia, 1985, p215.
3. Lekki, J.; Laskowsky, J. In *Colloid and Interface Sci.*, Vol.4, M.Kerker, Ed.; Academic Press, New York., 1976.
4. Hornsby, D.; Leja, J., Selective flotation and its surface chemical characteristics, *Surface Colloid Sci.*, 1982, Vol. p.217.
5. Conrad, S.H.; Wilson, J.L.; Mason, W.; Peplinski, W.J., *Water Resources Research*, 1992, Vol. 28, in press.
6. Wilson, J.L.; Conrad, S.H.; Mason, W.R.; Peplinski, W.; Hagan,E., R.S. Kerr Environmental Research Laboratory; *1990,EPA/600/690/004.*
7. Weyl, W.A., *J.Colloid Sci.*,1951, Vol.6, p389.
8. Usui, S.; Dasaki, H., *J. Colloid Interface Sci.*, 1978, Vol.65, p36.
9. Usui, S.; Dasaki, H., *J.Colloid Interface Sci.*, 1981, Vol. 81, p80.
10. Yoon, R.H.; Yordan,J.L., *J.Colloid Interface Sci.*, 1986, Vol.113, p.430.
11. Lee, C.Y.; McCammon, J.A., *J. Chem. Phys.* 1984, Vol. 80, p4448.
12. Williams, D.F., *Aggregation of Colloidal Particles at the Air-Water*

Interface; Ph.D. thesis, Department of Chemical Engineering, University of Washington, 1991.
13. Derjaguin, B.V.; Landau, L.D., Acta Phys. Chem URSS13, In *Theory of the Stability of Lyophobic Colloids*, Verwey, E.J.W.; Overbeek, J.T., Ed.; Elsevier, Amsterdam 1948.
14. Marcelja, S.; Radic, N., *Chem. Phys. Lett*, 1976, Vol. 42, p.129.
15. Christenson, H.K., *J. Dispersion Sci. Technol.*,1988, Vol. 9, p.171.
16. Dzyaloshinskii, L.E.; Lifshitz, E.M.; Pitaevski, L.P., *Adv. Phys.* 1961, Vol.10, p.165.
17. Claesson, P.M., *Prog. Colloid Polym. Sci.*, 1987, Vol.74, p.48.
18. Churaev, N.V.; Derjaguin, B.V., *J. Colloid Interface Sci*, Vol.103, p.542.
19. Laskowski. J.; Kitchener, J.A., *J. Colloid Interface Sci.*, 1969, Vol. 29, p.670.
20. Rabinovich, Ya. I.; Derjaguin, B.V., *Colloids Surf*, 1988, Vol. 30, p.243.
21. Derjaguin, B.V.; Racinovich, Ya. I.; Churaev, N.V., *Nature (London)*, 1978, Vol. 272, p.313.
22. Claesson, P.M.; Golander, C.G., *J. Colloid Interface Sci.*, 1987, Vol. 117, p.366.
23. Xu, Z.; Yoon, R.H., *J. Colloid Interface Sci.*, 1990, Vol.132, p.427.
24. Israelachvili, J.N., Faraday Discuss. *Chem. Soc.* 1978, Vol.65, p.20.
25. Israelachvili, J,N.; Pashley, R.M., *J. Colloid Interface Sci.*, 1984, Vol.98, p.500.
26. Claesson, P.M.; Blom.C.E.; Herder, P.C.; Ninham, B.W., *J. Colloid Interface Sci.*, 1986, Vol.114, p.235.
27. Pashley, R.M., *J.Colloid Interface Sci.*, 1981, Vol.83, p.531.
28. Christenson, H.K.; Claesson, P.M.; Pashley, R.M., *Proc. Indian Acad. Sci. Chem Sci.*, 1987, Vol. 98, p.379.
29. Mates, T.E.; Ring, T.A., *Colloids Surf.*, 1987, Vol.24, p.299.
30. Claesson, P.M.; Kitchener, J.A., *J. Colloid Interface Sci.*, 1969, Vol. 29, p.670.
31. Derjaguin, B.V.; Zorin, Z.M., *Zh.Fiz.Khim*, 1951, Vol. 29, p.1010.
32. Rapacchietta, A.V.; Neumann, A.W., *J. Colloid Interface Sci.*, 1977, Vol. 59, p.555.

RECEIVED March 20, 1992

Chapter 6

Colloid Remediation in Groundwater by Polyelectrolyte Capture

H. E. Nuttall, Jr.[1], S. Rao[1], R. Jain[1], R. Long[2], and Ines R. Triay[3]

[1]Department of Chemical and Nuclear Engineering, University of New Mexico, Albuquerque, NM 87131
[2]Department of Chemical Engineering, New Mexico State University, Las Cruces, NM 88003
[3]Los Alamos National Laboratory, Los Alamos, NM 87545

The presence of radioactive colloids (radiocolloids) in groundwater has been documented in several studies. There is significant evidence to indicate that these colloids may accelerate the transport of radioactive species in groundwater. Because field experiments are often fraught with uncertainties, colloid migration in groundwater is an area of active research and the role and existence of radiocolloids is being investigated. This paper describes an ongoing study to characterize groundwater colloids, to understand the geochemical factors affecting colloid transport in groundwater, and to develop an in-situ colloid remediation process. The colloids and suspended particulate matter used in this study were collected from a perched aquifer site (located at Los Alamos National Laboratory's Mortandad Canyon in northern New Mexico, USA) that has radiation levels several hundred times the natural background and where previous researchers have measured and reported the presence of radiocolloids containing plutonium and americium. At this site, radionuclides have spread over several kilometers. Inorganic colloids collected from water samples are characterized with respect to concentration, mineralogy, size distribution, electrophoretic mobility (zeta potential), and radioactivity levels. Presented are the methods used to investigate the physiochemical factors affecting colloid transport and the preliminary analytical results. Included below are a description of a colloid transport model and the corresponding computational code, water analyses, characterization of the inorganic colloids, and a conceptual description of a process for in-situ colloid remediation using the phenomenon of polyelectrolyte capture.

Several studies indicate that radioactive colloids are present in groundwater and can contribute to accelerated migration of radioactive species through the subsurface (1, 2,

0097–6156/92/0491–0071$06.00/0

3, 4,5). Others *(7,8)* have indicated the likely potential for facilitated transport of contaminants by groundwater colloids. Longworth and Ivanovich *(1)* have measured both the concentration of radiocolloids and the amount of radioactive species adsorbed on colloids which were collected from a variety of aquifers. The partitioning of natural series actinides has been measured and both uranium and thorium were found on the colloids. Adsorption of radionuclides onto organic colloids was greater than for inorganic colloids but both colloid types would be considered radiocolloids since they contain radionuclides. It has been suggested in the literature that subsurface transport of contaminants and transuranic elements can be due in part to colloid migration *(5,9)*. At Los Alamos National Laboratory (LANL), two waste sites, TA 21 DP West and Mortandad Canyon, have been monitored over many years for the presence and migration of radiation. Predicted actinide migration distances were on the order of a few millimeters; however, monitoring results showed that at the TA 21 DP West site *(6)* plutonium and americium have migrated vertically downward nearly 33 m and at Mortandad Canyon, plutonium and americium were detected over two miles from the source *(5)*. It should be noted that the presence of colloids at Mortandad Canyon does not prove that radiocolloids were the cause of the contaminant migration. Further studies are needed to verify the role of colloids at both the TA 21 DP West and the Mortandad Canyon sites. Several other radioactive sites have been monitored for radionuclide migration in groundwater. Monitoring data has shown actinide and/or radionuclide migration in excess of 100 meters in groundwater. Sites other than Mortandad Canyon and TA 21 DP West where radionuclides have migrated considerable distances include the radioactive waste burial sites at Maxey Flats in Kentucky *(10)*, the Idaho National Energy Laboratory—INEL *(9)* and at the Nevada Test Site—NTS *(4)*.

Field data from these sites underline the need to better understand and predict the role of radiocolloids in contaminant migration. To meet this objective, a joint project was initiated. The objective is to understand and predict colloid-contaminant migration in groundwater through colloid transport modeling, water sampling with colloid characterization, and laboratory experiments. Using this combined information, the plan is to develop an effective and scientifically based colloid immobilization strategy. At DOE's Mortandad Canyon site, we are investigating the potential role of radiocolloids in the migration of plutonium and americium. Our progress in understanding ground-water colloids and their transport is summarized in this paper. In the first section, the site modeling and colloid transport code is described. Next are studies characterizing colloids and groundwater from Mortandad Canyon. To date we have investigated the groundwater composition, colloid size distribution, composition, zeta potential, and radiation levels. This characterization information is used in both the modeling and in the design of laboratory experiments. Lastly, we describe in a conceptual overview our proposed in-situ colloid remediation technique using polymer induced capture.

Colloid Transport Modeling

The purpose of modeling is to better resolve the role of colloids in facilitated radionuclide migration. As a first step, we are modeling the Mortandad Canyon hydrology and solute transport. Next, for the same hydrology conditions, we will model colloid transport.

Colloid transport modeling is also coupled to the design and interpretation of laboratory column experiments. The transport modeling can be divided into two activities: 1) study of Mortandad Canyon hydrology with solute transport and 2) colloid transport. The TRACR3D (*11*) code, which is well established and recognized for this application, is being used to provide groundwater velocity profiles and baseline solute transport predictions. Results from the TRACR3D simulations of the hydrology and solute transport at Mortandad Canyon were presented in a recent M.S. thesis (*12*). The water velocity profiles calculated from the TRACR3D simulations will be used as input to the Colloid Transport Code, (CTC) (*13,14*) for colloid transport simulation studies. Using the velocity profiles calculated in TRACR3D and appropriate submodels describing colloid capture, CTC will be used to model and predict the extent of colloid migration at Mortandad Canyon.

To describe colloidal transport CTC, which is a new code, solves the coupled solute transport and population balance equations. The population balance (*15*) is a continuity equation for the number of particles and the dependent variable is number density of particles. Particle properties such as colloid size are treated as additional orthogonal axes. The CTC code has been tested on a number of standard problems and successfully solved the problem of particle-wall interactions in two-dimensional colloidal transport through saturated/unsaturated fractures (*14*). The population balance model and code are very generic in nature and can be used to model colloid transport in a wide range of applications. Submodels specific to applications can be added to the code by the user. In view of the general nature of the model formulation, CTC was designed to solve unsteady, nonlinear, coupled, second-order differential systems in up to four spatial axes. It is written in standard Fortran 77 and can operate on most UNIX/ VMS workstations as well as Cray computers under the UNICOS operating system. The code automatically discretizes all the spatial derivatives by user-selected finite differencing schemes thus converting all partial differential equations into sets of ordinary differential equations. Boundary conditions of the most general form can be specified by the user and are added to the system of ordinary differential equations. The resulting system of equations is solved using the robust and efficient LSODPK (*16*) solver package. Graphical output is best viewed using NCSA Image graphics software (*17*).

To date, preliminary modeling results for the site hydrology and solute transport for Mortandad Canyon (*12*) have been consistent with the literature, i.e., results showed that dissolved species transport acting under normal ionic equilibrium behavior would not account for the extent of radionuclide migration present at Mortandad Canyon (*5*). Non-site specific, colloid transport modeling studies have shown that colloids can migrate very rapidly through small fractures under certain idealized conditions. This fact may help explain observations of rapid colloid migration both in the laboratory and in field experiments. All of these results are preliminary and the issues of contaminant migration at Mortandad Canyon are continuing to be investigated. The next section describes the colloid sampling and characterization activities. Actual groundwater colloid properties (size, composition, mobility, and mineralogy) are essential for both colloid modeling and for the design and interpretation of laboratory experiments.

Colloid Sampling / Characterization

Water chemistry and colloid properties affect colloid migration and must be measured in order to predict colloid mobility. The majority of this work is being performed by the colloid characterization laboratory located at the University of New Mexico. Groundwater colloid samples used in this study were collected from Mortandad Canyon. Water samples were drawn from three properly cased monitoring wells (MCO-4B, -6B and -7A) in the lower portions of Mortandad Canyon from boreholes spaced approximately a quarter of a mile apart. Samples were collected at a depth of approximately 40 to 60 feet in the   perched aquifer. More information on the Mortandad Canyon site is available in a recent LANL report (18). Water analyses for the three well samples are reported in Table I. Note that for the samples in columns B and C the solids were acid digested while for column A similar water chemistry analyses were performed but without acid digestion by the Soil Water Air Testing laboratory at New Mexico State University (NMSU).

The size distributions of the larger colloids contained in the three unfiltered water samples were analyzed using the Coulter Electronics Ltd. Multisizer (range = 1.07 to 30 μm) and the size distribution information is presented in Table II. It can be seen that the mean size of the colloids is fairly consistent over the three samples.

The size distribution of the smaller colloids was analyzed using the Coulter Electronics Ltd N4MD Submicron Particle Analyzer (range = 3 to 3000 nm) and the results are given in Table III.

The electrophoretic mobility of the groundwater colloids was measured using a Coulter Electronics Ltd DELSA 440. The results showing that the colloids are negatively charged are presented in Table IV.

Additional analyses performed on the water sample from well MCO-6B include measurement of:

    (a) Total Suspended Solids (TSS) : Method 2540 D of Standard Methods (19).

    (b) Total Dissolved Solids (TDS) : Method 2540 C of Standard Methods.

    (c) Electrical Conductivity : Method 2510 B of Standard Methods.

and the results are presented in Table V.

    Characterization of the water samples from the three Mortandad Canyon wells indicates a number of factors. First ,the total suspended solids, TSS, values reported in Table V are sensitive to gravity settling indicating that the original samples contained large particulate matter in addition to colloids. The existence of large particles was also confirmed by SEM analyses. Particles as large as 43 μm were observed. Hence, the unagitated TSS analyses are likely to be more representative of actual colloid concentrations. Better water sampling techniques are called for in future sampling. The measured colloid sizes varied considerably due to the gravity settling effects and the presence of large particulates. The water is of low ionic strength with a pH of about 8.7. The Mortandad colloids are negatively charged.

Table I. Water Analysis of Samples from Mortandad Canyon Monitoring Wells

Sample		MCO-4B			MCO-6B			MCO-7A		
Element	Units	A	B	C	A	B	C	A	B	C
Ag	(μg/L)	-	<5	0.3	-	<5	1.3	-	< 5	0.4
Al	(mg/L)	<0.1	-	15	<0.1	-	113	0.1	-	57.4
As	(μg/L)	300	<40	15.1	430	<40	12.7	220	<40	15.8
B	(μg/L)	140	-	-	130	-	-	160	-	-
Ba	(μg/L)	-	190	337	-	690	1670	-	420	820
Be	(μg/L)	-	<1	2.1	-	4	8.3	-	3	4.7
Ca	(mg/L)	44.1	-	55.4	38.9	-	53	17.6	-	25
Cd	(μg/L)	-	<5	0.9	-	<5	0.7	-	<5	0.7
Cl	(mg/L)	-	-	< 0.5	-	-	34.4	-	-	28.1
CN	(mg/L)	-	0.01	0.041	-	<0.01	0.046	-	<0.01	0.026
Co	(μg/L)	<20	<20	-	<20	<20	-	<20	<20	-
Cr	(μg/L)	-	<10	17.3	-	30	22.5	-	20	28
Cu	(μg/L)	<20	10	16.5	<20	30	17	<20	30	21.2
Fe	(mg/L)	<0.05	-	-	< 0.05	-	-	0.05	-	-
Hg	(μg/L)	-	<1	<0.2	-	<1	<0.2	-	<1	0.2
K	(mg/L)	29.3	-	45.1	21.9	-	32.8	5.9	-	11.3
La	(mg/L)	<0.01	-	-	<0.01	-	-	<0.01	-	-
Mg	(mg/L)	4.7	-	5.66	6.9	-	10.2	3.8	-	5.78
Mn	(mg/L)	<0.02	-	0.518	<0.02	-	2.56	<0.02	-	1.62
Na	(mg/L)	148.3	-	209	185.9	-	278	71.1	-	112.6
Ni	(μg/L)	<50	<20	10.9	<50	<20	17.3	<50	30	20.3
NO3-N	(mg/L)	-	-	50.2	-	-	15	-	-	18.8
P	(mg/L)	-	-	0.361	-	-	0.876	-	-	0.924
Pb	(μg/L)	<100	<30	42.3	<100	70	163	120	50	94
Rb	(μg/L)	<50	-	-	<50	-	-	< 50	-	-
Sb	(μg/L)	-	<30	0.5	-	<30	<0.5	-	<30	<0.5
Se	(μg/L)	<50	<60	2.5	<50	<60	2.2	<50	<60	1
SiO2	(mg/L)	35.4	-	-	28	-	-	45.6	-	-
Sn	(μg/L)	-	<20	-	-	<20	-	-	<20	-
SO4	(mg/L)	-	-	46.5	-	-	54.9	-	-	22.9
Sr	(mg/L)	0.19	-	-	0.21	-	-	0.1	-	-
Sulfides	(mg/L)	-	2	-	-	1	-	-	1.6	-
Tl	(μg/L)	-	<40	0.4	-	<40	2.1	-	<40	0.8
V	(μg/L)	-	<10	171	-	30	155	-	40	147
Zn	(μg/L)	10	81	72	10	150	149	20	100	107
TDS	(mg/L)	600	-	712	760	-	834	300	-	220
pH		-	-	7.54	-	-	7.31	-	-	6.96
Cond.	μmho/cm	-	-	717	-	-	905	-	-	220

A=NM State Univ;   B=Los Alamos-IT;   C=Los Alamos-HSE-9

Table II. Size Analyses of the Groundwater Colloids

	MCO4B	MCO6B	MCO7A
Window (μm)	1.07-8.57	1.07-7.81	1.07-7.81
Number of Counts	572.1x103	580.2x103	735.1x103
Mean (μm)	1.867	1.779	1.654
Median (μm)	1.518	1.467	1.441
Mean/Median Ratio	1.230	1.213	1.148
Mode (μm)	1.115	1.115	1.115
Std. Dev. (μm)	0.998	0.891	0.670
Variance (μm2)	0.996	0.795	0.449
Skewness	2.789	2.868	3.086
Kurtosis	9.923	1.033x101	1.430x101

Table III. Size Analyses of the Colloids
(using Coulter Electronics Ltd, N4MD)

	MCO4B	MCO6B	MCO7A
Unimodal			
Mean (nm)	2280	657	573
95% Conf. Limits (nm)	2090-2480	626-688	548-598

Table IV. Electrophoretic Mobility Analyses of the Colloids
(using Coulter Electronics Ltd, DELSA 440, Units = μm.cm/V.s)

MCO4B2	MCO6B2	MCO7A2	
Mode	-1.43	-1.42	-0.71
Mean	-1.51	-1.45	-0.71
Std. Dev.	0.4655	0.40	0.414
Skewness	-0.234	-0.075	0.02

Table V. Water Characteristics (MCO-6B)

Property	Unagitated	Agitated
TDS (mg/l)	887	865
TSS (mg/l)	6	2384
pH (at 22.2oC)	8.7	8.7
Conductivity (mmhos/cm)	1,000	1,000

Chemical composition data for both the colloids and from a core sample were measured using a Scintag X-Ray Diffraction apparatus. Results suggested that the colloids consisted of approximately the same mineralogical composition as that of the soil matrix, i.e., quartz, feldspar and montmorillonite clays which are aluminosilicates with traces of rare earth metals. The grain size of the particles of the soil matrix was determined by dispersing the particles in deionized water followed by ultra sonification and size measurement using the Multisizer. The size range was 0.5 - 1.5 mm.

The concentration of radionuclides on or within the colloids is being measured and because of the low concentrations it is a difficult and expensive procedure. However, LANL has reported that the suspended solids in Mortandad Canyon water samples do contain a very low concentration of plutonium on the order of approximately 0.03 pCi/g (*19*).

In-Situ Colloid Remediation Strategy

An in-situ colloid remediation concept is being developed and tested. This concept involves treating a contaminated subsurface zone with a dilute polymer solution to create polymer-induced capture on the rock matrix. A schematic diagram of the process is illustrated in Figure 1. Figure 2 shows the chemical formula for the proposed poly-

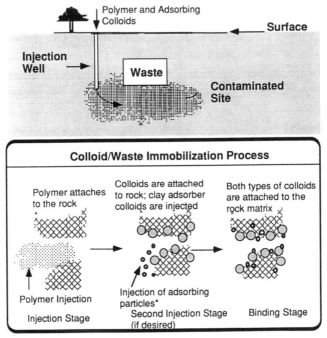

* adsorbing clay/zeolite colloids are added to remove dissolved radionuclides from solution

Figure 1. In-Situ Colloid Remediation Process.

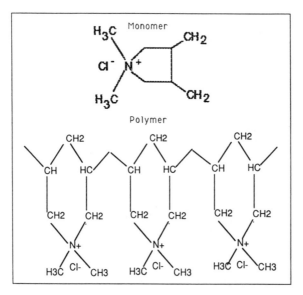

Figure 2. Proposed Polyelectrolyte for In-Situ Colloid Remediation Process

electrolyte, CATFLOC—poly diallyl dimethyl ammonium chloride. Polyelectrolyte induced flocculation has been used in municipal water treatment, and the theory is well developed (*21*). In our process, a polyelectrolyte solution is injected into the subsurface and the polymer is adsorbed onto the rock matrix in very low concentrations. The polyelectrolyte is characterized by two properties (i) charge density and (ii) molecular weight. The charge density with the polymer structure determines the extent of charge neutralization on the colloid surface when the polymer attaches to the surface of the colloid. Note that the charge on the polymer monomers is generally positive while the colloid surface is negatively charged. The higher the molecular weight and thus the greater the physical length of the polymer the more likely it is to form a bridge between the colloids and the rock matrix. The combination of these two properties of the polyelectrolyte lead to flocculation and attachment of colloids to the rock matrix. Ideally the addition of the polymer to a contaminated groundwater site would remove the radiocolloids. Addition of a natural clay type colloid (adsorber) would strongly adsorb the residual dissolved radionuclides and then polymer bridging (*22*) could attach these radioactive clay colloids along with natural radiocolloids onto the rock matrix.

Both batch and column type laboratory tests, are used in developing the in-situ colloid remediation process. Tasks include:
- determine efficiency of various polyelectrolytes for removing groundwater colloids.
- estimate polyelectrolyte dosage for given water and subsurface media characteristics.
- study effects of the following parameters,
    (i) water chemistry,
    (ii) colloid size and concentration,
    (iii)permeability of the medium, and
    (iv)aging and biodegradation.
- study mechanisms and kinetics of both the polymer attachment to the rock and colloid/rock interactions determine the extent of polymer spreading in a formation. The column studies are performed in small vertical polycarbonate tubes (0.8 cm x 17 cm) with selected packing, colloids, and water composition. The column and syringe pump system is illustrated schematically in Figure 3. Table VI shows the following matrix of column experiments now being performed.

Table VI. Colloid Remediation Development Experiments

Experiments	Packing	Dispersed	Continuous
1)	Quartz	Latex spheres	D.I.water
2)	Glassbeads	Latex spheres	D.I.water
3)	Core from site	Latex spheres	D.I. water
4)	Quartz	Colloidal dispersion	(site)
5)	Glass beads	Colloidal dispersion	(site)
6)	Core from site	Colloidal dispersion	(site)

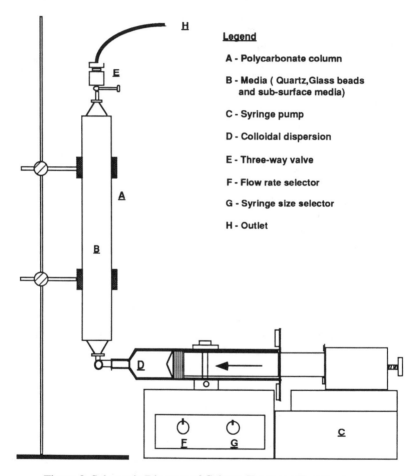

Figure 3. Schematic Diagram of Column Transport Test Equipment

Preliminary results showed that Mortandad colloids passed through packed columns of clean glass beads with little or no capture onto the surface of the glass beads. This finding further corroborates the zeta potential measurements which showed that the colloids are negatively charged.

Conclusions

The potential contamination of groundwater by radionuclides is of public concern. Several geologically separated nuclear sites have reported unusual and rapid migration of actinides in groundwater. In most of these accounts, radiocolloid migration is suggested as a possible or contributing mechanism. The purpose of this study is to better understand and clarify the role of colloids in groundwater transport of actinides. Our studies focus on the Mortandad Canyon site because of its close location and because colloids have been implicated in the migration of plutonium and americium at this site. This study combines site characterization modeling, groundwater/colloid sampling and characterization, and laboratory experiments. As a potential solution for the problem of colloid migration, we are developing an in-situ colloid remediation technique. Packed columns are used to investigate both colloid migration and colloid remediation on a laboratory scale. Transport modeling is being developed and applied to the Mortandad Canyon site to understand potential mechanisms for radionuclide migration. Colloid size, composition and electrokinetic analyses of radiocolloids from the site have been presented. Preliminary results indicate the presence of significant concentrations of negatively charged colloids in Mortandad Canyon groundwater. These colloids passed readily through columns of packed glass beads. The application of the polyelectrolyte CATFLOC appears promising for in-situ colloid remediation. Remediation results are still preliminary but are encouraging.

Acknowledgments

The authors thank Los Alamos National Laboratory for their assistance in providing water samples, analyses, and information about Mortandad Canyon. The research on which this publication is based was financed by the U.S. Department of Energy through the New Mexico Waste Management Education Consortium. A special thanks of appreciation is given to Cheryl Brozena for preparing this manuscript.

Literature Cited

(1) Longworth, G.; Ivanovich, M. Waste Management '90, Tucson, Arizona, **1990**.
(2) Travis, B. J.; Nuttall, H. E. Scientific Basis for Nuclear Waste Management VIII *44*, **1985** : 969-976.
(3) Nuttall, H. E. "Colloid Transport Update-Milestone R528";1989, LA-UR-87-3742.
(4) Buddemeier, R.W.; Hunt, J.R. *Appl. Geochemistry*, **198**, *3*(5) 535-548.
(5) Penrose, W. R., et al. *Environ. Sci. Technol.*, **1990**, *24*(2), 228-234.

(6) Nyhan, J.W.,; B.J. Drennon; W.V. Abeele; M.L. Wheeler; W.D. Purtymun; G. Trujillo; W. J. Herrara; J.W. Booth. Distribution of Plutonium and Americium Beneath a 33-yr-old Liquid Waste Disposal Site, *J. Environ. Qual.* 14(4), **1985**, 501-508.

(7) Ramsay, Radiochimica Acta, **1988,**44,167.

(8) McCarthy, John F.; Zachara, John M. *Environ. Sci. and Technol.*, **1989,** 23(5): 496-502.

(9) Rawson, S. A.; Walton,J. C.; Baca, R. G. Radiochimica acta, **1991,**52/53, 477-486.

(10) Meyer, G. Lewis. "Preliminary Data on the Occurrence of Transuranium Nuclides in the Environment at the Radioactive Waste Burial Site Maxey Flats, Kentucky"; EPA-520/3-75-021; 1976.

(11) Travis, B. J.; Birdsell, K. H., "TRAC3D: A Model of Flow and Transport in Porous Media: Model Description and User's Manual"; Los Alamos National Laboratory Manual, 1988, LA-11798-M.

(12) Vollick, Todd M.S. Thesis, New Mexico State University, 1991.

(13) Jain, Rohit M.S. Thesis, University of New Mexico, 1991.

(14) Fertelli, Yasemin M. M.S. Thesis, University of New Mexico, 1990.

(15) Randolph, A. D.; Larson, M. A., *Theory of Particulate Processes*, 2d ed.; Academic Press, 1988.

(16) Brown, P. N., and Hindmarsh, A. C. *SIAM J. Num. Anal.*, *23* (3), **1986**, 610-638.

(17) NCSA Software, Development-Image, 152 Computing Applications Bldg., 605 E. Springfield Ave., Champaign, IL 61820.

(18) Environmental Surveillance at Los Alamos During 1988, LA-11628-ENV, June, 1989.

(19) Stoker, kA. K.; W. D. Purtymun; S. G. McLin; M. N. Maes, Los Alamos Natonal Laboratory Report, LA-UR-91-1660, 1991.

(20) Standard Methods for the Examinaton of Water and Waste Water, 17 ed., American Public Health Association, John Wiley, 1989.

(21) Schwoyer, William L. K. "Polyelectrolytes for Water and Wastewater Treatment"; CRC Press Inc., 1981.

(22) Naper, D. H.; Academic Press, New York, 1983.

RECEIVED January 6, 1992

Inorganic Compounds

Chapter 7

Removal of Chromate from Aqueous Streams by Ultrafiltration and Precipitation

Edwin E. Tucker[1,2], Sherril D. Christian[1,2], John F. Scamehorn[2,3], Hirotaka Uchiyama[1,2], and Wen Guo[1,2]

[1]Department of Chemistry and Biochemistry, [2]Institute for Applied Surfactant Research, and [3]School of Chemical Engineering and Materials Science, University of Oklahoma, Norman, OK 73019

The present research seeks to establish an efficient and economic process for removing toxic anions such as chromate, arsenate, and selenate from aqueous streams without requiring addition of large quantities of chemicals. We have established a direct ultrafiltration process which is capable of removing over 99% of chromate ion (CrO_4^{2-}) from an aqueous feed stream. The ultrafiltration process using a cationic polyelectrolyte is also capable of recovering a large fraction of the input stream and producing it as permeate water of high quality under low operating pressures. The polyelectrolyte is separated from the toxic metal by a precipitation step and can be reused. The process should be particularly useful for treating ground water and wastewater streams contaminated with low concentrations of toxic metal anions.

Current technology applied for removal of toxic metal cations from aqueous streams consists primarily of precipitating the metals as hydroxides by addition of lime. Anionic materials such as chromate ion (CrO_4^{2-}) cannot be precipitated in this process (1). A prior step to reduce the chromium (VI), existing as chromate with a minus 2 charge, to chromium (III) which exists as the cation Cr^{3+} must be done before precipitation with lime. This reduction step might use either ferrous sulfate or an acidic sulfite reduction process. Once the chromium is in the +3 state, lime can be added to neutralize the solution and precipitate multivalent metal cations, including chromium, as the hydroxide. The entire treatment plan requires the addition of large quantities of sulfuric acid for pH adjustment and, subsequently, large quantities of lime (or other base) for neutralization and precipitation of the mixed metal hydroxides. The process produces a large volume of metal hydroxide sludge of low solids content to be disposed of (2). The costs for both chemicals and sludge removal are quite high because such large volumes are involved.

0097–6156/92/0491–0084$06.00/0
© 1992 American Chemical Society

We have developed, over the last several years, techniques for combined use of colloidal substances and ultrafiltration (UF) membranes for purification of aqueous streams containing both dissolved organic and inorganic (ionic) materials (*3-10*). Our objective has been to devise low-cost and highly efficient means of removing pollutants from dilute aqueous streams. These purification processes are characterized both by low energy requirements and low intensity chemical treatment of a target stream. The processes involve addition of a colloid, which may be either a surfactant or a polyelectrolyte, to an aqueous stream to sequester target materials. The solution mixture is then passed through an ultrafilter which retains the colloid-target material mixture while the permeate from the ultrafilter is water of substantially improved purity.

Membrane processes have been used previously for ion removal from aqueous streams. Reverse osmosis (RO) membranes can be used to remove large quantities of both anions and cations (*11*). However, water recovery (i.e., the fraction of the input stream which is produced as purified water) is generally quite low and the RO process requires very high pressures for operation. The use of UF membranes with a permanent charge has been applied in research on removal of oxyanions of arsenic and selenium. Rejections of divalent oxyanions as high as 95% were obtained with applied pressures as high as 80 psig (*1*).

Colloid Enhanced Ultrafiltration

Our ultrafiltration processes which utilize surfactant micelles as the added colloid are referred to as micellar-enhanced ultrafiltration (MEUF) processes since the micelles will be retained by a UF membrane with an appropriate molecular weight cut-off (MWCO). Surfactant micelles will solubilize organic solutes such as chlorinated aromatics and, if ionic surfactants are used, the micelles will attract multivalent ions of charge opposite that of the charged surfactant headgroup and thus these materials will be retained by the membrane.

Polyelectrolytes may be substituted for surfactants when waste streams containing primarily ionic material are to be purified. Polyelectrolytes such as sodium poly(styrenesulfonate) can be used to sequester multivalent ions of charge opposite that of the polymer and prevent passage through an UF membrane. We refer to such processes as polyelectrolyte-enhanced ultrafiltration or PEUF. Typically, polyelectrolytes of molecular weight range from 50,000 to 500,000 daltons are used.

One prior study which we have published is particularly pertinent to the present discussion of anion removal from waste streams (*8*). Table I gives a partial listing of data obtained using both dialysis and batch ultrafiltration cells with solutions containing the surfactant cetylpyridinium chloride (CPC) and sodium chromate. Chromate ions in this solution essentially replace the chloride counterions of the surfactant micelle and thus chromate is retained with the micelle as the solution is forced through the ultrafiltration membrane. Very low concentrations of chromate are observed in the permeate; up to 99.9% of the original chromate in the aqueous solution is kept in the retentate (i.e., rejected by the membrane) in this procedure.

The favorable exchange of chromate for chloride is essentially based on electrostatic attraction of the highly charged chromate ion to the positively charged pyridinium headgroups of the micelle.

Table I. Dialysis and Micellar-Enhanced UF Results for CPC-CrO_4^{2-}

NaCl Ret(M)[a]	CrO_4^{2-} Ret(M)	[CPC] Ret(M)	CrO_4^{2-} Per(M)[b]	Rejection %
	Dialysis	Results		
0.0	4.01e-3	0.104	5.50e-6	99.9
0.0	7.93e-6	0.104	2.36e-5	99.7
0.0	1.03e-2	0.104	5.47e-5	99.5
0.0	4.60e-3	0.054	1.02e-5	99.8
0.0	4.75e-3	0.025	2.31e-5	99.5
0.0	4.20e-3	0.010	3.36e-5	92.0
0.025	7.80e-4	0.100	2.00e-5	97.4
0.025	4.05e-3	0.100	1.41e-4	96.5
0.025	4.18e-3	0.050	2.80e-4	93.3
	Batch Cell	UF Results		
0.0	2.39e-2	0.119	4.60e-4	98.1
0.0	1.20e-2	0.120	6.20e-5	99.5
0.0	4.80e-3	0.119	5.80e-6	99.9
0.050	4.90e-3	0.122	2.80e-4	94.3

a. Ret=retentate; b. Per=permeate

The dialysis results refer to a method in which a chromate/CPC solution is placed on one side (retentate) of a membrane cell with pure water being placed on the other side (permeate). The membrane retains almost all the CPC in a 24 hour experiment. At equilibrium, both the retentate and permeate solutions are analyzed by UV spectroscopy for both CPC and chromate. Although the dialysis method is not useful for large-scale purification of aqueous streams, this technique offers a simple and accurate method for predicting separations which might be obtained in actual UF experiments (8,12).

Batch cell results were obtained with a 400 ml. stirred ultrafiltration cell. The feed CPC/chromate/NaCl solution is placed in the cell and filtered by applying 60 psig pressure from a nitrogen gas cylinder. Permeate and retentate analyses for CPC and chromate concentrations are performed at the end of the experiment. Rejection values for the data in Table I are defined as 100(1-[per]/[ret]) where [per] and [ret] refer to the analytical concentrations of chromate in the permeate and retentate samples, respectively.

The data in Table I show that the CPC micelles are extremely effective in removing chromate from an aqueous sample. One unexpected result is that the separation actually improves as the chromate concentration decreases in the feed mixture. This is unusual in a separation procedure. Other methods for removing ions from water, such as precipitation, become poorer as the concentration of the target ion decreases. A negative effect on the separation described above occurs on addition of NaCl. A theoretical model related to counterion binding in polyelectrolyte solutions (*13*) has been developed to correlate the above results (*8*).

Although the separation utilizing micelles is highly efficient, there is one minor problem involved. The micelle exists in equilibrium with the surfactant monomer and is thus not a covalently bonded 'polymer'. Consequently, as the feed solution is ultrafiltered, surfactant monomers 'leak' through the membrane at a concentration roughly equal to the critical micelle concentration. For the CPC surfactant this concentration is ca. 0.0009 M. The loss of surfactant monomer through the membrane leads to a small continuous depletion of the colloid material responsible for separation.

The loss of monomeric colloid which occurs with the use of surfactant micelles in UF separations may be avoided by substituting an appropriate polyelectrolyte for the surfactant in situations where inorganic (ionic) material is to be removed from an aqueous stream. In one demonstration of the use of polyelectrolyte (instead of surfactant) we have used the anionic polymer sodium poly(styrenesulfonate) in UF separation of Cu^{2+} from aqueous solution (*7*). Highly purified effluents were obtained with rejections of Cu(II) as large as 99.95% in these experiments. Added sodium chloride was again shown to have a negative effect on the quality of the separation. However, rejections of Cu(II) of better than 90% could be obtained as long as the concentration of added salt did not exceed the polystyrenesulfonate (PSS) concentration. One very important result of this study was that the efficiency of separation of the target ion increased substantially with decreasing Cu(II) concentration while the feed ratio of Cu(II) to PSS was kept constant. This is in accord with the previous indication of improved separation of CrO_4^{2-} with CPC micelles at lower CrO_4^{2-} concentrations (*8*).

To effect removal of CrO_4^{2-} or similar anions from aqueous streams using a polyelectrolyte it is necessary to use a cationic polymer which is analogous to the cationic micelle CPC. The cationic polymer poly(dimethyldiallyl ammonium chloride) or PDMDAAC has been used in the present experiments.

The measurements discussed to this point have involved small scale dialysis cells or 400 mL batch UF cells to establish the feasibility of the colloid enhanced UF methods. For utility in real-world applications it is necessary to scale-up these separation experiments to membrane systems with substantially larger volume capacities and membrane areas. Hollow fiber UF columns and spiral wound UF cartridges are examples of two types of UF units which are appropriate. Figure 1 gives a schematic of a recently constructed UF system which utilizes a hollow fiber column.

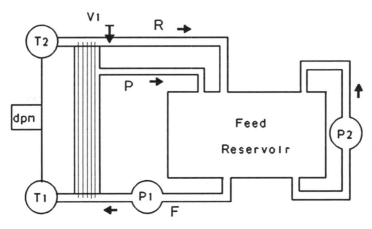

Figure 1.   Hollow Fiber Ultrafiltration Apparatus

The feed reservoir is a 5 gallon polyethylene tank. The tank contents are internally recirculated at a 300 gal/h rate by pump **P2** to prevent concentration gradients. Feed solution is forced into the hollow fibers by pump **P1**, which has an accurately controlled pumping rate up to 2 L/min. The fibers are represented by the thin vertical lines. Retentate exits at the top of the column through tube **R** and permeate exits the fibers horizontally, flowing out through the tube marked **P**. The column is a Harp 18 inch, polysulfone membrane hollow fiber cartridge with 2 ft^2 area and a 10K molecular weight cut-off (MWCO). **T1** and **T2** are pressure transducers measuring input and output (back) pressures, respectively, up to 50 psig for the column and displaying readings on the digital panel meter **(dpm)**. Variable volume splitting of feed into retentate and permeate fractions is accomplished by opening or closing valve **V1** which modifies column back pressure. Restriction of retentate flow by partially closing valve **V1** increases the fraction of feed exiting horizontally (through the fiber walls) to the column as permeate. The apparatus is operated in steady state mode with both retentate and permeate being recycled to the feed reservoir. Sampling points for feed, retentate, and permeate flow are not shown. Withdrawal of samples for analysis leads to a slight decrease in feed concentration over the course of an experiment. In addition to the composition of the feed solution, the two interdependent factors of feed pumping rate and retentate flow restriction may be modified to change separation conditions. Chromate analysis is made by direct UV measurement at 372 nm and, for very dilute chromate solutions, measurement at 542 nm using a diphenylcarbazide colorimetric reagent.

The ultrafiltration apparatus depicted in Figure 1 has been used in initial experiments to test the efficiency of removal of CrO_4^{2-} from aqueous streams using the quaternary amine polymer PDMDAAC. A polymer sample of nominal molecular weight 200,000 is first dissolved in distilled water and circulated through the hollow fiber column to remove short polymer chains and other contaminants which will permeate the 10K MWCO column. In these experiments we have used a polymer concentration of 0.001 M (on a monomer unit basis) and a sodium chromate concentration of 0.0001 M. This chromate concentration is ca. 5 ppm (as chromium) which is an appropriate level of chromate to be used for simulating dilute waste streams.

Table II gives results for one UF experiment with a polyelectrolyte/$[CrO_4^{2-}]$ concentration ratio of 10:1.

Table II. Hollow Fiber UF Results for PDMDAAC-Sodium Chromate Solution

CrO_4^{2-} Feed(M)	CrO_4^{2-} Ret(M)	CrO_4^{2-} Per(M)	Flow[a] L/min	Recov. %	Reject.[b] %	P in psi
9.76e-5	1.19e-4	2.3e-7	0.671	14.5	99.8	4.6
9.67e-5	1.33e-4	2.1e-7	0.718	25.5	99.8	6.2
9.54e-5	1.63e-4	1.6e-7	0.710	38.9	99.8	9.4
9.43e-5	1.81e-4	1.6e-7	0.720	46.5	99.8	12.8
9.22e-5	1.99e-4	1.8e-7	0.716	53.1	99.8	17.9
8.93e-5	2.09e-4	1.8e-7	0.723	54.8	99.8	24.1

Conditions: nominal flow 0.750 liter/min of feed solution with initial composition of 0.001M PDMDAAC and 0.0001M sodium chromate. Column input pressure varied from 4.6 psig initially to 24.1 psig as the permeate fraction was increased. (a) This is the combined (retentate + permeate) flow out of the hollow fiber cartridge in liter/min. The next column of figures gives the per cent of this flow which is permeate. (b) Feed-based rejection is defined as 100(1-[per]/[feed]) where [per] and [feed] are chromate concentrations in permeate and feed solutions, respectively.

The results of this experiment are impressive. Chromate concentration in the permeate has been reduced by over two orders of magnitude compared with the feed concentration. Permeate concentrations of CrO_4^{2-} of ca. 2e-7 M correspond to a Cr(VI) level of about 10 ppb compared with the feed concentration of 5000 ppb. Separation has been accomplished with quite low levels of applied pressure also. Permeate recovery of ca. 55% of the feed volume requires no more than 24 psig column input pressure.

Our previous studies with other systems have shown that the UF separations become poorer as the ionic strength of the feed solution is increased by adding NaCl. Separations with better than 90% rejection have been possible so long as the added salt was not present at concentration levels higher than the colloid concentration. We have obtained data for the PDMDAAC-chromate system with a high level of salt concentration. Table III presents data for which the salt concentration is 10 times higher than the polyelectrolyte concentration. The data in Table III show that there is still a substantial separation of chromate at this high salt concentration although it is clear that either lower salt levels or higher polyelectrolyte concentrations are necessary to preserve highly efficient separation.

Table III. Hollow Fiber UF Results for PDMDAAC-Na$_2$CrO$_4$-NaCl Solution

CrO$_4^{2-}$ Feed(M)	CrO$_4^{2-}$ Ret(M)	CrO$_4^{2-}$ Per(M)	Flow L/min	Recov. %	Reject.[a] %	P in psi
9.38e-5	9.89e-5	3.62e-5	0.292	13.0	61.4	2.1
9.33e-5	1.13e-4	3.55e-5	0.287	30.0	62.0	4.2
9.23e-5	1.35e-4	3.38e-5	0.289	45.0	63.4	10.1
9.03e-5	1.68e-4	3.16e-5	0.288	58.0	65.0	16.5
8.83e-5	1.82e-4	3.06e-5	0.290	65.0	65.4	22.1

Conditions: nominal flow 0.30 liter/min of feed with initial composition of 0.001M PDMDAAC, 0.0001 M sodium chromate and 0.01 M NaCl. Column input pressure varied from 2.1 psig initially to 22.1 psig as the permeate fraction was increased. (a) per cent rejection is defined as 100(1-[per]/[feed]) where [per] and [feed] are chromate concentrations in permeate and feed solutions, respectively.

Table IV gives results obtained with the hollow fiber UF apparatus for a polymer/chromate solution both with and without added salt at the level of polymer concentration. The separation is somewhat poorer with added salt but still shows chromate rejections of greater than 95% for this system. The separation improves very slightly as the recovery percentage increases. The same effect is evident from the data in Table III also. This probably reflects the fact that the membrane is transparent to NaCl, i.e., while the polymer and chromate concentrations increase in the retentate as permeate is removed, the NaCl concentration presumably does not.

Table IV. UF Results for 0.0056M PDMDAAC/0.0001M Na$_2$CrO$_4$

NaCl(M) Feed	CrO$_4^{2-}$ Feed(M)	CrO$_4^{2-}$ Ret(M)	CrO$_4^{2-}$ Per(M)	Recov. %	Reject. %	P in psi
0.0	1.18e-4	1.35e-4	1.7e-7	8.9	99.9	3.93
0.0	1.18e-4	1.83e-4	1.5e-7	35.0	99.9	9.95
0.0	1.17e-4	2.44e-4	1.5e-7	47.9	99.9	19.8
5.6e-3	1.17e-4	1.24e-4	5.94e-6	8.1	94.9	2.91
5.6e-3	1.16e-4	1.43e-4	5.80e-6	19.9	95.0	6.15
5.6e-3	1.17e-4	1.64e-4	4.84e-6	32.0	95.8	11.8
5.6e-3	1.16e-4	1.88e-4	4.10e-6	39.8	96.5	19.6
5.6e-3	1.16e-4	2.02e-4	3.84e-6	43.0	96.7	25.1

Conditions: nominal flow of 0.350 L/min of feed solution.

Spiral Wound Ultrafiltration Apparatus

The experimental results from the hollow fiber UF apparatus are very encouraging. However, the factors of an upper limit to transmembrane pressure of 25 psi (which essentially limits the possible permeate recovery percentage in this apparatus) and possible scale-up problems in going to larger membrane systems led us to construct another apparatus utilizing a spiral wound membrane cartridge of 5 ft^2 area. The spiral wound system is similar in concept to the hollow fiber apparatus. Specific differences were the use of components which could be interfaced to a small computer for better control and convenience. The spiral wound cartridge (Spectrum Industries, cellulose acetate, 20K MWCO) has an upper pressure limit of 100 psi. A remotely controlled gear pump, pressure transducers, and pneumatic sampling valves were interfaced with the computer.

Table V gives one set of results obtained with the spiral wound ultrafiltration apparatus. The feed-based rejection of chromate is quite good but is slightly poorer than that obtained in the smaller hollow fiber apparatus. As the fraction of the feed solution taken off as permeate increases, the rejection of chromate by the membrane becomes somewhat poorer. The total system flow decreases substantially in the course of this experiment because the input pressure is held constant as opposed to the hollow fiber experiments where the input volume flowrate was held constant.

Table V. Spiral Wound UF Results for PDMDAAC-Na$_2$CrO$_4$[a]

CrO$_4^{2-}$ Feed(M)	CrO$_4^{2-}$ Ret(M)	CrO$_4^{2-}$ Per(M)	Flow L/min	Recov. %	Reject. %	P in psi
4.44e-5	6.24e-5	2.65e-7	0.970	29.3	99.4	30.3
4.42e-5	6.78e-5	3.13e-7	0.826	34.7	99.3	30.5
4.42e-5	7.30e-5	3.72e-7	0.735	40.0	99.2	30.5
4.39e-5	8.01e-5	4.19e-7	0.633	44.7	99.1	30.3
4.36e-5	1.02e-4	6.38e-7	0.475	58.1	98.5	30.3
4.30e-5	1.31e-4	8.99e-7	0.393	68.4	97.9	30.3
4.10e-5	2.27e-4	2.13e-6	0.307	83.0	94.8	30.6

a. Feed solution is 0.00125 M polymer and 0.00005 M Na$_2$CrO$_4$

Separation/Regeneration of Polymer

The UF process which has been described has great promise for removal of chromate from aqueous streams and thus producing an effluent (permeate) which is extremely low in chromate concentration. However, the retentate material contains a concentrated solution of polymer and chromate which cannot readily be discarded. An efficient and economic process would consist of a further step to separate the polymer/chromate mixture to provide a compact chromate waste and a polyelectrolyte-rich stream which could be returned to the process. The most direct route toward freeing the chromate from the polyelectrolyte and producing a compact chromate waste appears to be a precipitation step. A suitable precipitating agent for chromate would further our objective of low intensity chemical treatment of the wastewater stream. The precipitant for chromate which might be least environmentally harmful in low concentration appears to be barium chloride. The chloride ion will reconstitute the polyelectrolyte with its chloride counterion and barium chromate may be separated as a solid waste. We consider the following set of simultaneous equations as a model for the precipitation of chromate:

$$K_{sp} = [Ba^{2+}] \, [CrO_4^{2-}] \quad \text{(solubility product for barium chromate)} \qquad (1)$$

$$\text{Added BaCl}_2 = [Ba^{2+}] + [BaCrO_4]\text{solid} \quad \text{(moles BaCl}_2 \text{ in 1 liter)} \qquad (2)$$

$$\text{Total CrO}_4^{2-} = [CrO_4^{2-}] + [BaCrO_4]\text{solid} \quad \text{(moles chromate in 1 liter)} \qquad (3)$$

$$[Cl^-] = 2 \text{ Added BaCl}_2 \qquad \text{(chloride molarity)} \qquad (4)$$

$$[Na^+] = 2 \text{ Total } CrO_4^{2-} \qquad \text{(sodium molarity)} \qquad (5)$$

$$2 [Ba^{2+}] + [Na^+] = 2 [CrO_4^{2-}] + [Cl^-] \qquad \text{(charge balance)} \qquad (6)$$

$$\% \ CrO_4^{2-} \text{ removed} = 100 \ [BaCrO_4]\text{solid/Total } CrO_4^{2-} \ (\% \text{ chromate pptd.}) \qquad (7)$$

$$\text{ppm } Ba^{2+} = 137360[Ba^{2+}] \qquad \text{(ppm barium in supernatant)} \qquad (8)$$

$$\text{Factor} = \text{Added } BaCl_2/\text{Total } CrO_4^{2-} \ \text{(mole ratio of barium to chromate)} \qquad (9)$$

These nine equations describe an aqueous solution mixture composed of certain amounts of sodium chromate and barium chloride. We have neglected hydrogen and hydroxyl ion concentrations although it is implicit that the system pH should be held at pH 7 or slightly above in order that the chromium present be almost totally in the form of chromate. Given the solubility product constant for barium chromate (ca. 1.2 e-10) and initial concentrations of barium chloride and sodium chromate, the first 8 equations can be solved simultaneously to determine the equilibrium quantities of chromate, barium, solid barium chromate, per cent chromate removed, and ppm of barium in the supernatant liquid. Adding equation 9 to the scheme gives us a mechanism for controlling the total barium chloride added to the system as a fraction of the total chromate present. For a given total chromate concentration in the system, we can solve the equation set multiple times while varying the barium chloride to sodium chromate ratio over a range from near zero to unity.

We will select a specific situation to illustrate the results. We assume that we have ultrafiltered a solution of PDMDAAC initially containing 0.0001 M chromate to a point at which the retentate solution has a volume which is 20% of the feed. At this point, 80% of the feed solution has been produced as permeate and virtually all of the chromate is contained in the retentate at a concentration which is now 0.0005 M. Assuming that the PDMDAAC-chromate solution is analogous to a sodium chromate solution, we begin to add barium chloride in a stepwise manner while solving the equation set (for each specific barium chloride addition) until a 1:1 ratio of added barium chloride to total chromate is reached. The most important variables here are the per cent chromate precipitated and the barium ion concentration in the supernatant. Figure 2 gives a plot of the increase in $[Ba^{2+}]$ as a function of the per cent chromate precipitated.

Figure 2 shows that the Ba^{2+} concentration in the supernatant liquid remains quite low until ca. 90% of the chromate has been precipitated. At this point, the barium is present at a level of 0.32 ppm in the liquid. It is important to note that Figure 2 indicates clearly why precipitation as a single process would not be as good for chromate removal as a single PEUF step. An overwhelming amount of $BaCl_2$ would be necessary for precipitation of chromate at the 99+ % level and, consequently, the level of Ba^{2+} in the effluent would rise to several hundred ppm.

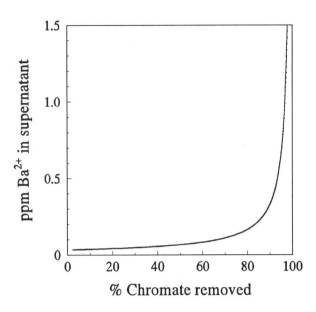

Figure 2. ppm Ba^{2+} in Supernatant as a function of % Chromate precipitated

The plotted results in Figure 2 are somewhat optimistic since we have ignored the presence of polyelectrolyte. In reality, the presence of polyelectrolyte should lead to a poorer separation (by precipitation) of chromate than predicted since much of the chromate in the real polyelectrolyte/chromate solution will be bound to the polymer and thus the chromate activity in solution will be lowered.

Two types of experiments have been performed to assess the effect of barium chloride precipitation of chromate in the presence of polymer. The first experiment was to determine the solubilizing power of the polymer with respect to solid barium chromate. Varying concentrations of polyelectrolyte in distilled water were contacted with an excess of solid barium chromate. After equilibration, the supernatant solutions were filtered and then analyzed for chromate. Figure 3 shows the resulting chromate concentration in solution as a function of the concentrations of two differently sized PDMDAAC polymers. Merquat 100 is of higher molecular weight (ca. 240K). The curve initially shows a steep rise in chromate concentration at very low PDMDAAC concentration and then levels off and appears to go through a shallow minimum for the Merquat 100 polymer. Presence of the polyelectrolyte increases the equilibrium concentration of chromate by more than one order of magnitude relative to the solubility of barium chromate in distilled water.

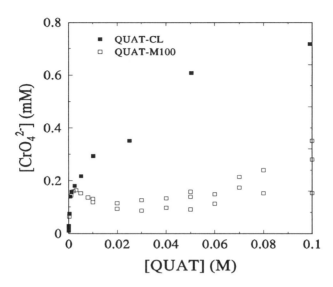

Figure 3.   Solubility of BaCrO4 vs. PDMDAAC concentration

The second set of experiments consisted of equilibrating various mixtures of polyelectrolyte, sodium chromate, and barium chloride.   The polyelectrolyte concentration range was from ca. 0.001M to ca. 0.12M.   At each specific polyelectrolyte concentration approximately equimolar concentrations of barium chloride and sodium chromate were added up to a maximum of ca. 0.001M each.

After equilibration, the solutions were filtered to remove precipitate and then analyzed for residual chromate.   Figure 4 shows two particular data sets of this type for PDMDAAC (Merquat 100) concentrations of 0.0167M and 0.0843M.   The convergence of the two lines indicates the point at which no $BaCrO_4$ will be precipitated in the polyelectrolyte-$BaCl_2$-Na_2CrO_4 mixture.   The lower sets of points which show the chromate concentration in solution (both free and polyelectrolyte-bound) are reasonably compatible with the solubility data in Figure 3 for the polyelectrolyte-$BaCrO_4$ mixture at the same polyelectrolyte concentrations.   The vertical difference between the two lines in Figure 4 is a direct measure of the amount of chromate precipitated as the barium salt.

Other Separation Considerations

We have concentrated on describing the specific features of removing chromate ion from ground water or waste streams.   The PEUF approach using a cationic polyelectrolyte should also be directly applicable for removal of other toxic multivalent oxyanions such as arsenate (AsO_4^{3-}) and selenate (SeO_4^{2-}).   Of course, these anions (and chromate) do not normally occur alone, with a single accompanying cation, in aqueous solution.   Further, the extent to which these toxic metals occur in the ionic form above is strongly affected by the solution pH. Two specific areas which should be discussed briefly are the pH effect on the specific ion form present and the complicating effects which might occur due to the presence of other ionic material in the solution under treatment.

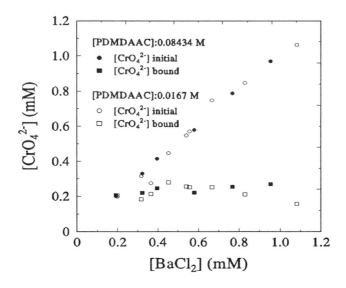

Figure 4. Chromate precipitation by Barium Chloride in presence of PDMDAAC

At solution pH values which are acidic (pH < 7) these anions will occur in protonated form (e.g., $HCrO_4^-$). The reduction of the anionic charge occasioned by protonation will impact the extent of interaction of the anion with the cationic polyelectrolyte; $HCrO_4^-$ would be expected to be much more weakly bound to the polyelectrolyte than CrO_4^{2-}. Another anion strongly affected by pH is arsenate. The ionization constants of the acid H_3AsO_4 are such that very little of the species AsO_4^{3-} will exist in aqueous solution except at quite high pH. However, model calculations show that the species $HAsO_4^{2-}$ (which should bind quite strongly to the polyelectrolyte) is present as more than 92% of the total arsenic at a pH of 8. It is clear that pH may be an important process variable and probably should be kept within a range between pH 7 and pH 10 for best results.

The presence of other ions in the process stream can cause interference in two major ways. These ions contribute to the ionic strength of the solution and, as we have previously discussed for the case of added NaCl, can reduce the efficiency of the PEUF process when present in excess concentration. Of course, if we are interested in treating a waste stream containing at least several ppm of the oxyanions above, then the concentration of a number of toxic metal cations such as lead (Pb^{2+}) will be low. Otherwise, precipitates such as lead chromate ($PbCrO_4$) would have already been formed. Other cations at high concentration could interfere due to ionic strength effects and these effects will need to be considered. It is, of course, quite feasible to use a prior UF step utilizing an <u>anionic</u> polyelectrolyte (7) to reduce the concentration of multivalent cations (e.g., Ca^{2+} or Cu^{2+}) in the incoming stream.

The second way in which other ions may interfere is by direct competition for sites on the polyelectrolyte and with the precipitant. For example, sulfate ion (SO_4^{2-}) will also be bound by the cationic PDMDAAC and precipitated by the addition of barium chloride to the retentate stream. Polyelectrolyte concentration in the process will need to be based on the <u>total</u> quantity of multivalent anions present which interact strongly with the PDMDAAC. Pretreatment of the incoming stream would also have to be considered if large quantities of relatively nontoxic multivalent anions occur with small quantities of chromate or arsenate or selenate.

Summary

We have developed a novel combination of ultrafiltration, using a polyelectrolyte to sequester toxic anions, and a precipitation process to both recover the anions as a solid waste and regenerate the polyelectrolyte for reuse. The method is highly efficient and works extremely well for dilute aqueous streams. Our technique also may have economic advantages in terms of both the low chemical costs involved and the greatly reduced solid waste volume anticipated (relative to the combined acidic reduction and hydroxide precipitation process for toxic anions). Our process does not add a salt burden (i.e., increased ionic strength) to the treated stream. Further research is necessary to determine appropriate operating parameters for optimum results but the technique appears quite promising for use in economic removal of low level toxic anion contamination in aqueous streams.

Acknowledgements

Sample quantities of the quaternary ammonium polymers Cat-Floc Cl and Merquat 100 were provided by Calgon Corp. (Division of Merck & Co., Inc.). Additionally, we wish to express our appreciation for support of this work through research grants from the U. S. Environmental Protection Agency (R-817450-01-0), the U. S. Department of Energy (DE-FG05-84ER13678), and the Oklahoma Center for the Advancement of Science and Technology.

Literature Cited

1. Bhattacharyya, D.; Moffitt, M.; Grieves, R. B. *Sep. Sci. Technol.* **1978**, *13*, 449.
2. Shapiro, N. I.; Liu, H.; Santo, J. E.; Fruman, D.; Darvin, C.; Baranski J.; Mihalik, K. in *Chemistry in Water Reuse*, Cooper, W. J., Ed.; Ann Arbor Science, Ann Arbor, MI, 1981, vol. 1, ch. 13, p. 281.

3. Christian, S. D.; Scamehorn, J. F.; Tucker, E. E.; O'Rear, E. A.; Harwell, J. H. in *CRC Handbook on Removal of Heavy Metals from Water*, in press.
4. Christian, S. D.; Tucker, E. E.; Scamehorn, J. F. *American Environmental Laboratory*, **1990**, *2(#1)*, 13.
5. Lee, B. H.; Christian, S. D.; Tucker, E. E.; Scamehorn J. F., *Langmuir*, **1990**, *6*, 230.
6. Christian, S. D.; Tucker, E. E.; Scamehorn J. F.; Sasaki, K. J.; Lee, B. H. *Langmuir*, **1989**, *5*, 876.
7. Sasaki K. J.; Burnett, S. L.; Christian, S. D.; Tucker, E. E.; Scamehorn J. F. *Langmuir*, **1989**, *5*, 363.
8. Christian, S. D.; Bhat, S. N.; Tucker, E. E.; Scamehorn J. F.; El-Sayed, D. A. *AIChe Journal*, **1988**, *34 (2)*, 189.
9. Bhat, S. N.; Smith, G. A.; Tucker, E. E.; Christian, S. D.; Smith, W.; Scamehorn, J. F. *I & EC Research*, **1987**, *26*, 1217.
10. Smith, G. A.; Christian, S. D.; Tucker, E. E.; Scamehorn J. F. in *Ordered Media in Chemical Separations*, Hinze, W. L.; Armstrong, D. W., Eds.; ACS Symposium Series #342, American Chemical Society: Washington, DC, 1987, pp. 184-198.
11. Kosarek, L. J. in *Chemistry in Water Reuse*, Cooper, W. J., Ed.; Ann Arbor Science, Ann Arbor, MI, 1981, vol. 1, ch. 12, p. 261.
12. Scamehorn, J. F.; Ellington, R. T.; Christian, S. D.; Penney, B. W.; Dunn, R. O.; Bhat, S. N. *AICHE Symp. Ser.*, **1986**, *82*, 48.
13. Oosawa, F. *Polyelectrolytes*, Marcel Dekker: New York, NY, 1971.

RECEIVED March 20, 1992

Chapter 8

Potential for Bacterial Remediation of Waste Sites Containing Selenium or Lead

L. L. Barton[1], F. A. Fekete[1], E. V. Marietta[1], H. E. Nuttall, Jr.[2], and R. Jain[2]

[1]Laboratory of Microbial Chemistry, Department of Biology and [2]Department of Chemical and Nuclear Engineering, University of New Mexico, Albuquerque, NM 87131

Under appropriate conditions, certain bacterial strains can reduce selenate or selenite to colloidal Se and Pb II can be transformed to a Pb-containing colloid. The red Se colloids have a diameter of 250 to 350 nm while the gray-black Pb colloids are 170 to 180 nm in diameter. Like the Se colloids, Pb colloids are spherical, of uniform density and not subject to disruption by sonic or surfactant treatments. Negative zeta potentials were observed for both the Pb and Se colloids across the pH range from 5.5 to 8.5. We propose the detoxification of wastes containing Se or Pb through bacterial metabolism to produce colloids. Complete removal of these elements from the environment is dependent on the implementation of systems appropriate for colloid recovery.

Bacteria have numerous systems for processing compounds which are toxic to higher forms of life. These cellular activities contributing to the removal of toxic metal species from the environment may include reduction to a less toxic form, or binding to cellular structures. Specific enzymes are produced by bacteria for the reduction of Cr VI to Cr III (*1*), Hg II to Hg° (*2*) and U VI to U IV (*3*). Extracellular polymers of bacteria exhibit an affinity, with considerable specificity, for metal cations including Cd, Co, Ni, Mn, Zn, Pb and Cu (*4*). The peptidoglycan in cell walls of *Bacillus megaterium* and *Micrococcus lysodeikticus*, two Gram-positive bacteria, binds Cd and Sr (*5*) while the outer membrane of *Escherichia coli*, a Gram-negative bacterium, binds Sr, Ni, Mn and Pb (*6*). Precipitates of heavy metal sulfides may result from the generation of hydrogen sulfide from *Desulfovibrio* and related anaerobic bacteria (*7*). The precipitation of U, Pb and Cd by the formation of insoluble phosphates has been reported to be enzyme dependent and to occur on the surface of *Citrobacter* sp. (*8*). The deposition of U, Ra and Cs within cells of *Pseudomonas aeruginosa* (*9*) or the formation of electron dense deposits of Pb by *P.*

0097–6156/92/0491–0099$06.00/0

maltophilia (R. Blake II, T. Zocco and L.L. Barton, unpublished reports) would be another example of a process which could remove specific toxic metals from the environment. The bioremediation of environments containing toxic metals has focused primarily on the use of ponds or streams. Precipitation of toxic metals from mining operations by the construction of impoundments have been proposed and would rely on activities of native bacteria (*10*). A wetland system for metal removal from acid mine drainage has been proposed and it would rely on the activity of sulfidogenic bacteria to produce insoluble heavy metal sulfides (*11*). Chelation resins constructed from dead *Bacillus* sp. have been proposed for removal of Cd, Cr, Cu, Hg, Ni, Pb, U, and Zn in fixed beds, fluid beds or dispersed beds (*12*). While these reports suggest the usefulness of bacteria for detoxification of environments, there is clearly a need to pursue alternate applications for the use of bacteria in bioremediation of toxic waste sites.

We report the use of two different bacterial isolates to reduce selenate or selenite to colloids containing Se and to transform Pb II to a Pb-containing colloid. Physical characteristics of these metallo-colloids have been conducted as a prerequisite to establishment of a system for the removal of these metal-containing colloids from aqueous environments.

Experimental Design for the Study of Metal Colloids

Growth of Bacteria. *Pseudomonas mesophilica* and *P. maltophilia* were initially isolated from soil on media containing 10 mM levels of Cr VI, Zn II and Pb II. Characteristics of these bacteria will be published elsewhere. The bacteria were grown in a broth medium containing 0.5% Bacto-Tryptone (Difco Co., Detroit, MI), 0.25% yeast extract (Difco Co.) and 0.1% D-glucose. For colloid formation, the bacterial cultures were grown in 100 ml medium placed in 300 ml baffled flasks with the addition of sodium selenite, sodium selenate or lead acetate to give a final concentration of 1.0 mM Se or Pb. The inoculated culture flasks were placed on a shaker adjusted to 200 rpm with a constant temperature of 32°C.

Preparation of Colloids. After incubation of the bacterial cultures for 5 days, the Se or Pb colloids were collected by centrifugation at 11,500 rpm for 2 min. The pellets, containing Pb or Se colloids, were resuspended in sterile saline, 0.85% NaCl solution, and the pellets were again collected by centrifugation. This saline wash was repeated until a total of three washes were used. For zeta potential measurements, washed colloids were resuspended in the following buffers: 0.1 M bis(2-Hydroxyethyl)iminotris(Hydroxymethyl)methane (Bis-Tris)-HCl, pH 5.5; 0.1 M N-2-hydroxyethylpiperazine-N'-2 ethanesulfonic acid (Hepes)-HCl, pH 7.0; and 0.1 M Hepes-HCl, pH 8.5. To measure the size of colloids, washed colloids were resuspended in distilled water. Removal of cellular material from the metallo-colloids was achieved by resuspending the colloids in 2 ml of a 1% sodium dodecylsulfate (SDS) solution and heating to 80°C. The SDS-treated colloids were washed in saline three times and resuspended in HEPES buffer, pH 7.0. Chemically produced colloids resulted from reacting equal volumes of 0.1 M sodium selenite with 0.1 M hydroxylamine and the resulting red colloid was washed three times in saline.

Colloid measurements. Analysis of colloids to determine size employed a Multi-sizer (Coulter Electronics Limited, Atlanta, GA.) or a Coulter Model N4MD Sub-Micron Particle Analyzer. The Multisizer is a multichannel particle analyzer which uses the electrical impedance method to provide a particle size distribution analysis using up to 256 size channels over a user-specified range. An orifice of 50 um diameter was used for obtaining an aliquot from the test material suspended in ISOTON II balanced electrolyte solution obtained from Coulter Electronics Limited. The size of the aperture tube allows the measurement of particles in the nominal diameter range of 1.07 to 30 um. The N4MD apparatus provides particle size analysis within the diameter range of 3 to 3000 nm by the measurement of sample diffusion coefficient by photon correlation spectroscopy.

The zeta potentials for colloids were established with a DELSA 440 instrument (Coulter Electronics Limited) with Doppler electrophoretic light scattering analyzer. Simultaneous laser Doppler spectra from four different angles (8.6°, 17.1°, 25.6° and 34.2°) were obtained and an average of these values were used to calculate the zeta potential. Results from the DELSA 440 were acquired and analyzed on an IBM-PC using Coulter's interfacing software.

Characteristics of Bacterially Produced Colloids

Se Colloids. The production of elemental selenium from selenite or selenate by bacteria has been the topic of several reports. Small, fibrillar selenium granules were demonstrated in the cytoplasm of *Pseudomonas* sp., *Aeromonas* sp., and *Flavobacterium* sp. (*13*). In *Escherichia coli* elemental selenium granules were observed in the cell wall (*14*) and large internal selenium granules were reported in *Wolinella succinogenes* (*15*) and in *Pseudomonas* sp. (Blake, Zocco and Barton, unpublished results). We have observed from microscopic examination that there was a loss of cell integrity of *P. mesophilica* growing aerobically in selenate or selenite with the appearance of red selenium in the culture. The absence of cells in the 5-day cultures provided an excellent opportunity to collect selenium colloids for study.

The Se colloids produced by *P. mesophilica* or hydroxylamine reduction of selenite existed in two sizes: the minimal colloid unit and as colloid aggregates (see Table I) An appreciable reduction in size of the native Se colloid aggregrate was observed following a 15 sec. exposure to sonication. The only minimal colloid to be reduced in size following sonication was the one produced by bacterial reduction of selenite. The diameter of the Se colloid produced from selenate reduction was unchanged by sonication; however, the Se aggregrate was completely dispersed by the mild sonic treatment. Similarily, the aggregates of Se colloids produced by chemical reaction or by bacterial reduction of selenite were reduced in size following sonic treatment. The size distribution of native and sonic treated Se colloids is graphically presented in Figure 1. While both graphic and tabular data are generated with each colloid size measurement, we find that the tabular information is a preferred method for expression.

When examining the relative abundance of these colloids in the reaction mixtures, colloid aggregates from chemical reaction or bacterial reduction were more abundant than the individual colloid. Only 9% of all the colloids produced from selenite reduction by *P. mesophilica* were of the minimal size while only 18% of the

Table I. Size of Selenium Colloids Produced
by Inorganic Reaction and by *P. mesophilica*

Source of Colloid	Native Colloid		Sonicated Colloid	
	Diameter (SD)[*]	Abundance as %	Diameter (SD)	Abundance as %
Chemical reaction:				
minimum unit	433 (150)	4%	315 (92)	8%
aggregrates	4,740 (1,400)	96%	2,610 (600)	92%
Reduction of selenite by bacteria:				
minimum unit	409 (41)	9%	247 (76)	36%
aggregrates	5,280 (720)	91%	653 (110)	64%
Reduction of selenate by bacteria:				
minimum unit	312 (42)	18%	357 (99)	100%
aggregrates	4,340 (610)	82%		

[*] Mean diameter is given in nm and (SD) refers to standard deviation.

native colloids produced from selenate were monodispersed colloids. In all cases, sonication resulted in a greater abundance of smaller colloids; however, the bacterially produced colloids were more readily dispersed than those from chemical reactions.

An additional difference between the colloids produced by chemical reaction and bacterial reduction was noted in chemical stability. Colloids produced by bacteria contained red amorphous selenium which was stable as this elemental form for several months. The chemically produced Se colloid was initially red but it became gray within a few days. It is well accepted that red amorphous selenium is relatively unstable and will gradually be converted to the more stable gray form (16). This suggests that the biologically produced Se colloid is stabilized by compounds produced by *P. mesophilica*.

Further characterization of these Se colloids was accomplished by measurement of zeta potentials. Typical results of analysis at the four degrees of measurement are presented in Figure 2. The mean zeta potential values and the standard deviations are shown in Table II for the colloids generated by the reductive process. It is apparent that Se colloids produced by bacteria were markedly different from the chemically produced Se colloids (Table II). Measurement at pH values of 5.5, 7.0 and 8.5 revealed considerable charge on the Se colloids. While electrophoretic mobility of bacteria have received considerable study (17), zeta potentials for bacteria and bacterial-derived colloids are not readily available.

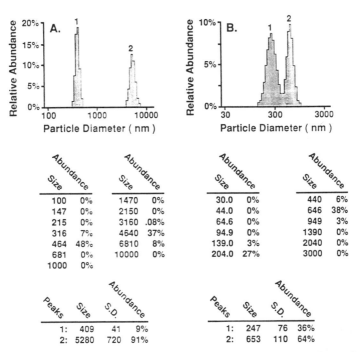

Figure 1. Measurement of colloids produced from reduction of selenite by bacteria. A. Native colloids; B. Sonicated colloids. Tabulated data are for each respective set of measurements.

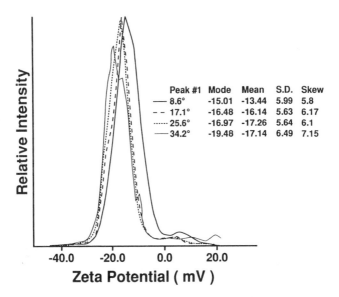

Figure 2. Data for zeta potential measurements with the selenium colloid tested at pH 5.5.

Table II. Zeta Potentials for Se Colloids and Bacteria

Source of Colloid	Values at Specified pH Levels		
	pH 8.5 mean (SD)*	pH 7.0 mean (SD)	pH 5.5 mean (SD)
Chemical reaction	+4.84 (2.26)	-40.0 (4.38)	+3.34 (9.26)
Reduction of selenite by P. mesophilica	-10.13 (17.8)	-5.78 (18.3)	-16.85 (5.92)
Reduction of selenate by P. mesophilica	-2.51 (11.2)	-3.24 (13.2)	-7.68 (6.98)
Cells of P. mesophilica	-11.65 (11.6)	+6.09 (12.0)	-0.54 (7.7)

* Mean values are given in mvolts and (SD) refers to the standard deviation.

Pb Colloids. The production of extracellular deposits of Pb have been reported for a bacterium which had been isolated from a toxic waste site and has been identified as a *Pseudomonas* (Blake, Zocco and Barton; unpublished results). Similarily, the bacterium which we isolated produced Pb colloids and was identified as *P. maltophilia*. Pb concentration in the medium was decreased as the cell density increased and, when tested by inductive coupled plasma analysis, Pb was found concentrated in the gray-black granules. The zeta potential of the Pb colloids was strongly negative and of greater negative charge than the cells of *P. maltophilia* (Table III). The diameter of the Pb colloids was 175 nm or about one-half the size of the selenium granules previously discussed. While these Pb colloids remain to be rigorously characterized, the charge on the colloid leads us to suggest that a system can be devised to collect the colloids from nature.

Colloid Recovery. In many instances it is sufficient to detoxify the environment of Se and Pb by forming insoluble compounds; however, in other cases it will be necessary to completely remove these metallo-compounds. In the removal of Se or Pb colloids, processes could be designed which would build on other technologies. One approach for the recovery of Se colloids has been proposed and involves stream-fed lagoons to allow for the sedimentation of Se compounds (*18*). Alternately, one could apply some of the procedures which have been employed for immobilization of cells and include the following: (i) adsorption by ionic binding onto a support material,

Table III. Characteristics of Pb-containing Colloids
Produced by *P. maltophilia*

Characteristic	Colloids	Cells
Zeta potential, mV.		
at pH 5.5	-42.04 (16.0)	-15.96 (4.95)
at pH 7	-35.28 (7.23)	-22.50 (5.88)
Diameter, nm		
at pH 7.0	175 (21.8)	Not Determined

(ii) covalent coupling established by addition of cross-linking reagents, and (iii) addition of matrix material to account for gel entrapment or microencapsulation. Systems employed in wastewater treatment plants could also prove effective in removal of the highly charged Se or Pb colloids through flocculation or charge neutralization procedures. Clearly the design of such a system for colloid recovery will be dependent on a series of physical and chemical parameters of the site requiring detoxification.

Other Metallo-colloids Which Could Be Produced by Bacteria

While our study has focused on colloid production by transformation of Se or Pb, micro-bioremediation by bacteria can be considered as a broader application. The transformation of metal salts to the elemental form by bacteria may be limited by the bioenergetic activity of the cells. Table IV lists reactions which could produce metal-containing colloids with reducing activity provided by NADH or NADPH coupled reactions. The half-cell reaction for oxidation of NADH or NADPH is -340 mV at pH 7.0 and would provide the energy for the reduction of the metals. In addition to having a favorable thermodynamic reaction, there must also be enzymes present to catalyze the specific redox reaction. It may be fruitful to examine other bacterial isolates for the capability to produce metal-containing colloids in addition to Pb and Se.

Concluding Remarks

Bacterial transformation of toxic metal salts to the base element would provide an excellent detoxification approach for many toxic wastes. From our results it is apparent that Se in the form of selenate or selenite may be reduced by bacteria to colloidal Se. The transformation of Pb II to an unidentified compound is significant because the Pb is immobilized and, therefore, the environment is detoxified. The Pb

Table IV. Reduction Potentials for Reactions Generating Metals
to the Elemental State

Reaction	Standard Reduction Potentials, mV (19)
Pb^{2+} + $2e^-$ $\rightleftarrows$ Pb^o	-126
Sb_2O_3 + $6H^+$ + $6e^-$ $\rightleftarrows$ $2Sb^o$ + $3H_2O$	+144
Bi^{3+} + $3e^-$ $\rightleftarrows$ Bi^o	+277
Te^{4+} + $4e^-$ $\rightleftarrows$ Te^o	+564
Ag^+ + e^- $\rightleftarrows$ Ag^o	+799
Hg^{2+} + $2e^-$ $\rightleftarrows$ Hg^o	+852
Au^{3+} + $3e^-$ $\rightleftarrows$ Au^o	+1445

compound also displays colloidal properties. Both the Se and Pb compounds produced by the bacteria readily settle out of solution because of the aggregation of the
of these biocolloids. We consider the surfaces of these colloids important relative to
interfacial activities and envision that processes can be developed to exploit
aggregration phemonema in removal of Se and Pb colloids. The process of
immobilization of toxic compounds through formation of complexes by bacteria is
a developing field of bioremediation.

Acknowledgments

The research on which this report is based was partially financed by the U.S.Depart-
ment of Energy through the New Mexico Waste Management Education. Also, F.A.
Fekete was the recipient of a grant for sabbatical leave from Colby College,
Waterville, ME.

Literature Cited

1. Ishibashi, Y; Cervantes, C; Silver, S. *Appl. Environ. Microbiol.* **1990**, *56*, pp.
    2268-2272.
2. Silver, S.; Misra, T.H. *Ann. Rev. Microbiol.* **1988**, *42*, pp. 717-743.
3. Lovley, D.R.; Phillips, E.J.P. *Nature.* **1991**, *350*, pp. 413-416.
4. Geesey, G.G.; Jang, L. In *Metal Ions and Bacteria*; Beveridge, T.J.; Doyle,
    R.J., Eds.; John Wiley & Sons: New York, NY, **1989**; pp. 325-358.

5. Marquis, R.E.; Mayzel, K.; Cartensen, E.L. *Can. J. Microbiol.* **1976**, *22*, pp. 975-982.
6. Hoyle, B.D.; Beveridge, T.J. *Appl. Environ. Microbiol.* **1983**, *46*, pp. 749-752.
7. Jackson, T.A. *Environ. Geol.* **1978**, *2*, 173-189.
8. Macaskie, L.E.; Dean, A.C. *Enzyme Microbiol. Technol.* **1987**, *9*, pp. 2-4.
9. Strandberg, G.W.; Shumate II, S.E.; Parrott, Jr., J.R.; North, S.E. In *Environmental Speciation and Monitoring Needs for Trace Metal-Containing Substances from Energy-Related Processes*; U.S. Department of Commerce, National Bureau of Standards: Washington, D.C., **1981**; pp. 27-35.
10. Gale, N.L. In *Workshop on Biotechnology for the Mining, Metal-Refining and Fossis Fuel Processing Industries*; Ehrlich, H.L; Holmes, D.S, Eds.; Biotechnol. Bioeng. Symp. No. 16; Wiley: New York, NY, **1986**; pp. 171-182.
11. Emerick, J.C.; Cooper, D.J. In *Proc. 90th National Western Mining Conf.*, Colorado Mining Association: Denver, CO **1987**.
12. Brierley, C.L.;Brierley, J.A.; Davidson, M.S. In *Metal Ions and Bacteria*; Beveridge, T.J; Doyle, R.J, Eds.; John Wiley & Sons: New York, NY, **1989**; pp. 359-382.
13. Silverberg, B.A.; Wong, P.T.S; Chau, Y.K. *Arch. Microbiol.* **1976**, *107*, pp. 1-6.
14. Gerrard, T.L; Telford, J.N; Williams, H.H. *J. Bacteriol.* **1974**, *119*, pp. 1057-1060.
15. Tomei, F.A; Barton, L.L.; Lemanski, C.L.; Siebring, R.J.; Zocco, T.G. In *Abstracts of the Annual Meeting of American Society for Microbiology.* **1989**, p. 362.
16. Zingaro, R.A.; Cooper, W.C. In *Selenium*; Van Nostrand and Reinhold Co.: New York, NY, **1974**; pp. 1-30.
17. Collins, Y.C.; Stotzky G. In *Metal Ions and Bacteria*; Beveridge, T.J.; Doyle, R.J., Eds; John Wiley & Sons: New York, NY, **1989**; pp. 31-90.
18. Baldwin, R.A.; Stauter, J.C.; Kauffman, J.W.; Laughlin, W.C. In United States Patent number, 4,519,913, **1985**.
19. Moeller, T. In *Inorganic Chemistry*; John Wiley & Sons: New York, NY, **1952**; pp. 286-290.

RECEIVED December 18, 1991

Chapter 9

Heap Leaching as a Solvent-Extraction Technique for Remediation of Metals-Contaminated Soils

A. T. Hanson, Z. Samani, B. Dwyer, and R. Jacquez

Department of Civil, Agricultural, and Geological Engineering, New Mexico State University, Las Cruces, NM 88003

EPA has recently initiated a renewed effort to devise technologies and treatment systems to remediate heavy metal contaminated soils on-site without generating significant wastes for off-site disposal. Heap leaching, a technique used extensively in the mining industry, has been investigated by the authors as a method for the remediation of heavy metals contaminated soils. In this paper the technical and economical feasibility of heap leaching is discussed and compared with other solvent extraction treatment techniques. Work to-date has been in laboratory scale columns similar to the columns used in the mining industry for preliminary mining studies. The authors have completed lab scale work with Cr (VI) and are currently in the midst of a study of Pb contaminated soil.

Until recently, efforts in improving waste handling techniques have been devoted to (1) reducing the generation of waste at its source, (2) monitoring stored waste to record its possible migration, and (3) developing better methods for emplacement to mitigate waste migration. Presently, there is a great urgency at the national level for the development of viable technologies for remediation of sites containing heavy metals. Numerous techniques have been proposed for the remediation of the vadose zone. However, one might question the use of the term "remediation" in describing many techniques. To refer to in-situ vitrification as soil remediation may be considered as equivalent to referring to death by electrocution as rehabilitation of a criminal. It is suggested that remediation should imply that the soil being treated must not be destroyed and the site must be returned to useful service at the end of the treatment. In the past the emphasis has been on fixing in-organic contaminants in-place or in excavating contaminated soils and re-burying them in a secure landfill. These solutions to problems with toxic inorganics provide only a temporary solution to the problem, since the inorganics do not degrade (1). Consequently, tremendous efforts have and are being made to devise technologies and treatment systems which remediate contaminated soil on-site, primarily because of increasing costs and the potential for long-term environmental impairment and liabilities (2).

0097–6156/92/0491–0108$06.00/0

Extraction Technologies

Solvent extraction of the metal from the soil is a technique which has received considerable attention lately, and it appears to have great potential. There are

three basic solvent extraction technologies currently being used in the mining industry which have potential to be used in the remediation of metals contaminated soils; vat leaching, solution mining, and heap leaching. Vat leaching and solution mining have received considerable attention in the waste treatment industry, but heap leaching has been largely ignored. This is particularly interesting since, in the mining industry, the first techniques of choice are solution mining and heap leaching. Vat leaching, the more complex and expensive technique, is only considered if heap leaching is not a viable alternative. The mining techniques and their hazardous waste treatment corollaries are described below.

The vat leaching technologies being used in soil remediation are referred to as "soil washing", and can be divided into two classes. Only one of the two classes is a solvent extraction technique. The first technique referred to as "soil washing" is similar to "ore enrichment" in the mining industry, and is actually a fines separation procedure. The technique would be more appropriately called "fines separation", since clay and silt material is washed out of a contaminated soil, separated from the liquid washing stream, and collected (*3*). The contamination is frequently associated with the large surface area in the fine fraction in the soil. This process removes the contaminant from the bulk of the soil. Figure 1 is the flow scheme for a commercially available process of this type marketed as the Lurgi Deconterra process. The fines separation systems have excellent process control and, the soil being confined in a reactor provides opportunity for excellent quality control and quality assurance (QC/QA). However, this technique has a number of disadvantages. The mixing, and jigging operations require energy, and the equipment requires a high level of operational skill. Probably the major disadvantage of this technology is that, although the soil washing reduces the volume of soil to be treated, it does not completely remediate the soil. The remaining contaminated fines must still be treated. In order to more fully research soil washing as a remediation technique, the EPA designed and now operates a mobile Volume Reduction Unit (VRU) for soil washing. The VRU is a pilot-scale mobile system for washing soil contaminated with a variety of heavy metals and organic contaminants. According to a US EPA report (*4*), soil washing has good to excellent applicability for cleaning soils with a sandy/gravely matrix, but has a moderate to marginal application for silty/clay soils.

The second technique referred to as "soil washing" is closely related to the vat leaching process used in the mining industry. The contaminated soil is excavated, and placed in an agitated vessel with an extraction solution. When the soil is sufficiently clean the solids are separated from the liquid and the liquid is processed or reused. Figure 2 shows a schematic of this technology. Unlike fines separation discussed above, the contaminating material is actually removed from the soil phase and is introduced into the liquid phase. The

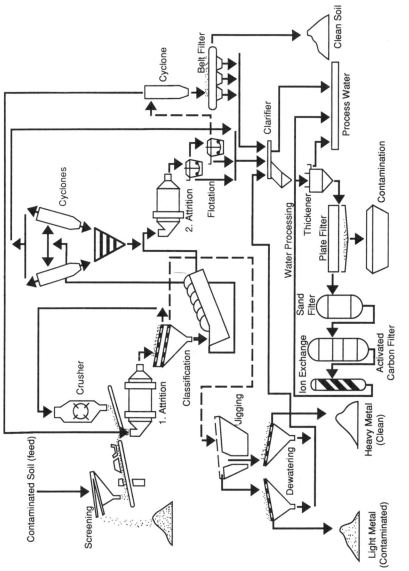

Figure 1.   Fines reduction unit built by a Dutch firm and marketed under the name Lurgi Deconterra Process.

contaminant must then be removed from the liquid phase. This is a common characteristic of all solvent extraction treatment processes. Like fines separation, this reactor based technology has good process control and QC/QA assessment, but is energy intensive and requires a high level of operational skills. This technology operates with solid:liquid ratios of between 1:10 and 1:70, and appears to be restricted to soils with fines concentrations of less than 30 percent (fines being defined as particles smaller than 60 μm).

Soil flushing is a technology for in-situ extraction of toxic materials, and it is very similar to "solution mining" used in the mining industry. Figure 3 illustrates the components of soil flushing. An extraction fluid is applied to unexcavated, undisturbed soil at ground level or at a depth, and the solution passes through the contaminated soil. At the base of the contaminated soil zone, the flushing fluid is recovered at the groundwater table using subsurface drainage pipes, trenches, wells or well points (5). The advantage of this system is the ability to treat large amounts of material without handling the material. The disadvantages are: the soil is non-isotropic and non-homogeneous, posing potential preferential flow problems. There is the potential for groundwater contamination. QC/QA assessment is difficult and the extraction fluid may cause the soils to swell and plug the aquifer. In the uranium mining industry, where this technology has been used extensively, groundwater contamination has been a serious problem. Tweeton (6) notes that field testing has shown that complete restoration, as defined by the state regulatory agencies, of solution mining sites has not been achieved with any reasonable degree of aquifer flushing. A number of others have also expressed concern over the restoration of aquifer materials contaminated with extraction solvent from solution mining operations (7-10). All of the expressed concerns combine to make the future for this treatment technique fairly bleak under most conditions.

The third solvent extraction technique, heap leaching, has been largely ignored as a hazardous waste treatment technique. Modern day precious metal heap leaching technology has developed over the past decade. In the heap leaching process, metal bearing ore of interest is excavated and mounded on a pad. The metals are removed by passing extraction fluid through the ore using some type of a liquid distribution system. Traditionally a simple sprinkler system was used, but recently drip irrigation systems have been employed (11). The extraction fluid is collected in a pregnant solution pit and the pregnant solution is processed to remove the metal of interest. This process is shown schematically in Figure 4. The advantages of this remediation technology are: the method is operationally simple, QC/QA assessment is practical, the method is not energy intensive, the liquid:solids ratio used is low, the method can be performed on-site on a large scale, the method is applicable to heavy clay soils, the method provides a permanent solution and eliminates the need for long term monitoring. Disadvantages are multiple handling of the contaminated material, and process control is difficult if one needs to adjust temperature and impossible if one would like to modify the pressure in the heap. The precious metals industry has recently introduced a process improvement, in which geothermal resources are used to heat the leach pile and accelerate the leaching process (11).

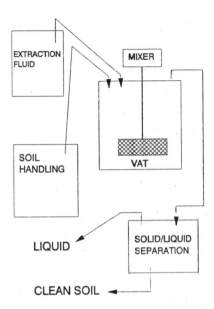

Figure 2.   A schematic of vat leaching, also called soil washing, in which solvent is used to extract contaminant from soil in a mixed reaction vessel.

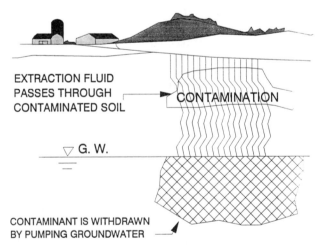

Figure 3.   Solution Mining, also called soil flushing, an in situ solvent extraction process.

Comparison of Metals Mining and Soil Treatment.

Considering the wealth of information available on heap leaching in the mining industry, it might appear that the outcome of heap leaching of toxic metals could be pre-determined. However, there are a number of major differences between the mining industry and the hazardous waste industry. The immediately obvious differences are physical and geochemical in nature. The metals chemistry and the relationship of the metals to the solid matrix in mining, is quite different than the typical hazardous metal contaminated soil. In the mining industry the metals of interest are intimately associated with the solid matrix, and this solid matrix consists of particles larger than 10 to 25 millimeter in diameter. In contrast, in a contaminated soil, the metal is adsorbed onto the surface of the contaminated soil particles. The soil particles generally have a size of less than .5 mm. The intimate nature of the metals association with the mineral matrix may significantly affect the dissolution time of the metal when compared with the dissolution time of a metal from the surface of a soil particle. The size of the soil particles will affect the extraction fluid application rates and the distribution of the extraction fluid in the leaching pile. In addition to the systems physical differences and hydrodynamic differences, the metals of interest are generally different. In the mining industry the metals of concern are gold, silver, and copper, while the metals of interest in the hazardous waste area are chromium, uranium, plutonium, lead, cadmium, etc.

In addition to the obvious physical and chemical differences, the motivational forces are also different. The mining industry is extracting metal from a rock matrix until an economic constraint is reached. In the hazardous waste industry one must continue to extract contaminant until a regulatory treatment goal is achieved.

Metal Extraction Technologies

Inspite of the fact that the overall systems are different, there is much that the hazardous waste industry can learn from the mining industries experience in removing metals from rock matrix using extraction fluids. Table I contains a list of metals which have been leached, the extraction technique which was used, and the extraction fluids which were successful.

During the past ten years, heap leaching has developed into an efficient method of mining oxidized gold, silver, and copper ores. It has provided an efficient way to extract metals from both small, shallow deposits, and large, low-grade, disseminated deposits. The first commercial application of heap leaching occurred in the late 1960's by the Carlin Gold Mining Company in northern Nevada. Cortez Gold Mines started the first large scale operation in the early 1970's by leaching two million tons of marginal grade ore (*19*). Heap leaching operations are now common, and range in size from small precious metals operations to huge copper extraction operations. Some of the copper leaching heaps in south western New Mexico cover many acres and are in excess of 240 feet high.

Table I. Metals which have been successfully mined using solvent extraction
(*12-18*)

Metal	Process	Extraction Fluid	Comments
Lithium (Li)	Vat	H_2O after roasting @ 900 °C	
Beryllium (Be)	Vat	H_2SO_4 then H_2O	
Sodium (Na)			
Magnesium (Mg)	Vat	SO_2 & Temp.	
Aluminum (Al)	Vat	Boiling 25% HCl	
Vanadium (V)	Vat	Boiling H_2SO_4	
Chromium (Cr)	Heap	Acid	Lab
Manganese (Mn)	Heap	SO_2 5% by wt	Lab
Cobalt (Co)	Heap	H_2SO_4 pH = 1.5-3.0	Cu by-product
Nickel (Ni)	Vat	Heat, Pressure, Mixing NH_4, H_2SO_4, HCl, & NO_3	
Copper (Cu)	Heap	H_2SO_4 pH = 1.5-3.0	Commercial
Zinc (Zn)	Vat	Chlorine & O_2 Heat and Pressure	
Arsenic (As)	Heap	Alkaline	U Tailings
Selenium (Se)	Heap	NaCN	Au by-product
Zirconium (Zr)	Vat	Dilute HCl	
Niobium (Nb)	Vat	Acid Boil, Alkaline	
Molybdenum (Mo)	Heap	3.03 % NaOCH	Lab
Silver (Ag)	Heap	NaCN	Commercial
Cadmium (Cd)	Vat	H_2SO_4 & Mixing	
Tin (Sn)	Vat	Roasting, H_2SO_4 & NaCl	
Hafnium (Hf)	Vat	Dilute HCl	
Tantalum (Ta)	Vat	Acid Boil, Alkaline	
Tungsten (W)	Heap	HCl	Lab
Gold (Au)	Heap	NaCN	Commercial
Mercury (Hg)	Heap	NaCN	Au by-product
Lead (Pb)	Vat	FeCl, Heat, pH=0.3	
Uranium (U)	Solution/Heap	CO_2	Commercial
	Heap	Alkaline	Tailings

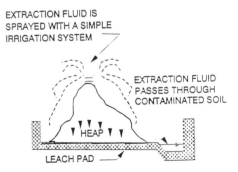

EXTRACTION FLUID IS SPRAYED WITH A SIMPLE IRRIGATION SYSTEM

EXTRACTION FLUID PASSES THROUGH CONTAMINATED SOIL

HEAP

LEACH PAD

Figure 4.   Heap Leaching; the simplest of the solvent extraction processes.

By the mid 1970's, heap leaching technology was being used with low-grade, clayey deposits. This improvement, known as agglomeration-heap leaching, was driven by an increase in exploration for low-grade deposits as the price of gold increased dramatically. Before agglomeration-leaching, many of the deposits discovered could not be processed using existing heap leaching techniques because the clay fines impeded uniform percolation of the leaching solution through the ore heaps. Current practice is to agglomerate the fines using 10 lbs of portland cement per ton of ore.

By 1986 the results of the technological improvements of the 70's and 80's were evident, as gold production from heap leaching had increased to over 30 percent of total U.S. gold production, up from an estimated six percent in 1979. Heap leaching also accounts for approximately 30% of the total copper mined in the United States.

The metal mining industry also shows that there is great potential in using this technology to meet regulatory treatment standards. A removal of 70 to 80 percent of the gold and silver in the matrix is typical, but removals of up to 95 percent are possible (11). The Round Mountain Mine in Central Nevada produces a leached ore with approximately 30 ppb of gold remaining in the matrix (20). This represents a 95 percent recovery. In the copper mining industry, 70 to 80 percent of the copper is typically recovered from low grade ores. These facts combine to make the study of heap leaching interesting and potentially very useful in the remediation of metals contaminated soils.

Laboratory Experiments

Three soils, which span the spectrum of soil characteristics common in the arid southwest, were selected and analyzed in accordance with ASTM D422 (21). The soils characteristics are shown in Table II.

Table II. Classification of Soils

Soil	% Sand	% Silt	% clay	ASTM Classification
Pajarito	85	8	7	Sand (S)
Harkey	62	22	16	Sandy Loam (SL)
Belen	22	36	42	Clay (C)

The soils have a low organic content (0.3 - 0.7 percent). The clays in these soils have been verified using X-ray diffraction. The saturated soil hydraulic conductivity of each of the three soils shown in Table III were determined in accordance with Designation E-13 (21).

Table III. Soil Hydraulic Conductivities

Soil	Soil Hydraulic Conductivity (cm/hour)
Pajarito (Sand)	9.7
Harkey (Sandy Loam)	4.8
Belen (Heavy Clay)	1
White Sands Gypsum	1

Prior to each test, the soils were contaminated with Cr(VI). A calculated and weighed quantity of potassium dichromate was dissolved in a volume of water equal to 15% of the volume necessary to saturate the soil volume being contaminated. The soil was spread in a thin layer. The potassium dichromate solution was broadcast evenly over the thin soil layer, the soil was allowed to air dry, the dry soil was loaded into the soil column, and the columns were leached under varying experimental conditions.

The laboratory scale Cr (VI) experiments were conducted in soil columns that were 7 feet high and 3.5 inches in diameter. The details of this work are available in Dywer (22). The results shown in Figure 5 demonstrate the effect of application rate on a heavy clay soil (Belen) contaminated with Cr(VI). These results are typical of those observed for other soils. Figure 5 shows that within five hours over 90% of the Cr(VI) added to the soils was removed, and in less than 30 hours over 99 % of the Cr (VI) was removed. The laboratory scale column study showed similar results for all three low organic soils from the arid southwest. The effect on leaching efficiency of soil type, hydraulic loading rate, intermittent saturation and drainage, ionic strength of extraction solution, and the effect of short term aging were all tested. Under all test conditions the number of bed volumes of extraction fluid (water) required to remediate the soil was the same (22).

During the column leaching studies, the Harkey, Belen, and White Sands soils were used to study the effect of agglomeration (described below) on the leaching process. The agglomeration is a technique used in the mining industry to enhance the leaching efficiency in heavy clay soils with low permeability. In the column studies there was little benefit in the agglomeration. However, in the field scale heap, where the liquid flow is not constrained by the column, there may be some benefit. The soils were contaminated as usual, then agglomerated in a standard concrete mixer. The agglomeration was accomplished by mixing the dry soil with Type II Portland cement (10 lbs. cement/ 2000 lbs. soil) until thoroughly blended, and wetting the soil/cement mixture, as it was mixing, with a mist of tap water (10 to 15% moisture content) until the soil agglomerated to 1/4" to 1/2" soil balls. The agglomerated soils were then leached as usual. The effect of agglomeration on the efficiency of leaching the heavy clay soils was also tested. The agglomeration did not decrease either the time required to treat the soil or the amount of water required to treat the soil. The time required for treatment of the agglomerated soil was nearly identical to the time

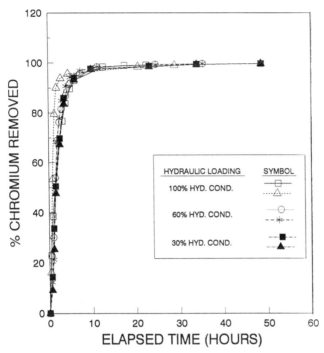

Figure 5.   The effect of hydraulic loading rate on leaching efficiency, when leaching Cr(VI) from Belen soil.

required to treat the un-agglomerated soil. The volume of water, however, was increased by a factor of 7.

The authors are currently investigating the extraction fluids suitable for the removal of lead from contaminated soils. Table IV contains the extraction solutions which have been tested at this point. The results in Table IV are based on small scale laboratory batch shaker studies. The most promising of the extraction solutions will be tested in the pilot scale columns. It is clear from the results in Table IV, that heap leaching has potential for treating soils which are contaminated with lead. Recent work by Peters and Shem[23] also support these findings. Using EDTA in a similar shaker test, Peters and Shem reported a 60-70 percent removal of lead from soil contaminated with lead nitrate.

The mere fact that something works is not sufficient information to design the system. If one is to design these systems it is critical that a number of question be addressed in transferring this technology from the mining industry to the hazardous waste industry. Parameters which are of interest if one is to design these systems are; the required removal time, the uniformity of the removal process, and the required volume of leaching agent under different conditions. For instance, Figure 6 indicates that regardless of the leaching rate 90 percent of the chromium contamination was removed from the soil column in 0.2 bed volumes or less. Figure 7 indicates that there is a

Table IV. Extraction Solutions for lead

LEACHING AGENT	SOIL	LEAD REMOVAL PERCENTAGES
Deionized Water	Laboratory Spiked	0.8%
Ferric Chloride	Laboratory Spiked	17.8%
Citric Acid	Laboratory Spiked	49.6%
Hydrochloric Acid	Laboratory Spiked	1.5%
Sodium Hydroxide	Laboratory Spiked	NR
SDBA	Laboratory Spiked	0.1%
SDS	Laboratory Spiked	0%
Sodium Cyanide	Laboratory Spiked	NR
DMSA	Laboratory Spiked	NR
CDTA	Laboratory Spiked	44%
CDTA	Laboratory Spiked *	86.3%
CDTA	Superfund	44%
CDTA	Superfund *	84%, 15.2%
EDTA	Laboratory Spiked	98%
EDTA	Laboratory Spiked *	NR
EDTA	Superfund	NR
EDTA	Superfund *	NR

NR = Results not yet received from lab.
* = Three day sequential leach. Fresh leaching agent added daily.

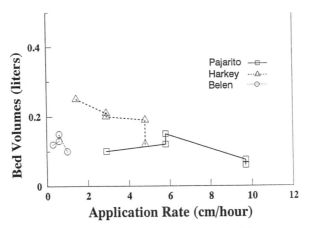

Figure 6.   Leachate generated in removing 90 percent of chromium from the three soils tested.

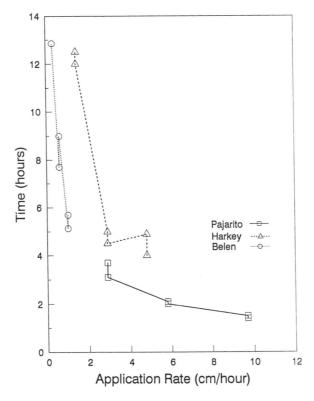

Figure 7.   Time required to remove 90 percent of chromium from a column of soil.

relationship between soil type, hydraulic loading rate, and time required to leach a specified height of soil column. While the work reported here has resulted in uniform removal of heavy metal from contaminated soils, the heap leaching of metals under field conditions may be different due to non-homogeneity of the soil, water channeling, geometry of the heap, and non-uniformity of extraction fluid application. Unfortunately this information is for columns and not for heaps. Similar information for heap leaches would make it possible to predict how large of a pad design one would need to accomplish the clean up of a contaminated site at a desired rate. This information is currently being developed for soils contaminated with Cr(VI) and Pb.

Acknowledgments

The research on which this publication is based was financed in part by the U.S. Department of Energy, through the New Mexico Waste-management Education and Research Consortium (WERC).

Literature Cited

1. Raghavan, R.; Wolf, G.; Wheeler, F. Technologies applicable for the remediation of contaminated soil at superfund radiation sites. Third international conference on new frontiers for hazardous waste management, Pittsburgh, Pennsylvania, 1989.
2. Castle, C. et al. Research and Development of a Soil Washing System for Use at Superfund Sites. Sixth Annual Conference on Management of Uncontrolled Hazardous Waste Sites, pp. 453-455.
3. Lauch, R. Evaluation of treatment technologies for contaminated soil and debris. Risk reduction engineering laboratory, EPA, Cincinnati, Ohio, 1989.
4. Rubin, B., et al. U.S. EPA's Mobile Volume Reduction Unit for Soil Washing. Superfund 90, Proceedings of the 11th National Conference, 1990, pp. 761-764.
5. Kunz, M.E.; Gee, J. R. Bench and pilot scale studies for metals and organics removal from CERCLA site soils. Superfund,89 ,proc. of the 10th national conference, 1989.
6. Tweeton, D. Restoring Groundwater Quality Following In Situ Leaching. In Situ Mining Research, Proceedings. Bureau of Mines Technology Transfer Seminar, Denver CO, Aug. 5, 1981.
7. Deutsh, W.,; Eary, L., Martin, S., McLaurine, S. Use of Sodium Sulfide to Restore Aquifers Subject to In Situ Leaching of Uranium Ore Deposits. Proceedings of the Association of Groundwater Scientists and Engineers. Western Regional Groundwater Conference, Reno, Nevada, Jan. 1985.
8. Canterford, J. H. Hydrometallurgy: Winning Metals With Water. *Chemical Engineering*. Vol. *92*, No. 22, pp. 41-48.
9. Charbeneau, R. J. *Groundwater Restoration with In Situ Uranium Leach Mining. Groundwater Contamination*; National Academy Press: Washington, DC., Feb., 1984.
10. Yelderman, J. C.,; Durler D. L. Detection, Correction, and Restoration of Mining Solutions in a Confined Aquifer at an In Situ Uranium Mine. Proceedings of the Third National Symposium on Aquifer Restoration and Groundwater Monitoring, National Water Well Assoc., Worthington, OH. 1983.

11. Trexler, D.,; Flynn T., Hendrix, J. L. Heap Leaching. Geothermal Heating Council Bulletin, Summer, 1990.

12. Bhappu, R. B.,; Reynolds, D. H. *Studies on Hypochlorite Leaching of Molybdenite.* Bureau of Mines: 1963.

13. Carnell, W. G.,; Choppin. *Plutonium Chemistry.* American Chemical Society: Washington, D.C., 1983.

14. Industrial Process Profiles for Environmental Use. U.S. Environmental Protection Agency, Industrial Environmental Research Lab., Cincinnati,OH., 1980.

15. Jeffers, T. H.,; Harvey, M. R. Cobalt Recovery from Copper Leach Solution. Bureau of Mines: 1985.

16. Kilau, H. W.,; Shah, I. S. *Preventing Chromium Leaching from Waste Slag Exposed to Simulated Acid Precipitation.* Bureau of Mines: 1990.

17. Pahlman, J. E. *Leaching of Domestic Manganese Ores with Dissolved SO.* Bureau of Mines: 1990.

18. Sandberg, R. G.,; Staker, W. L. *Calcium Sulfide Precipitation of Mercury During Cyanide Leaching of Gold Ores.* Bureau of Mines: 1984.

19. Thorstad, L.E. How Heap Leaching Changed the West, World Investment News, A Pacific Regency Publications, Vancouver, B.C., Feb. 1987.

20. York, D. A.,; Aamodt P. L. Remediation of Contaminated Soil Using Heap Leaching mining Technology. Western Regional Symposium on Mining and Mineral Processing Wastes, Berkley, CA, May 30-June 1, 1990.

21. Earth Manual, A Water Resources Technical Publication. U.S. Department of Interior: Bureau of Reclamation, 1974.

22. Dwyer, B. *Remediation of Hexavalent Chromium Contaminated Soils by Heap Leaching,* Masters Thesis, New Mexico State University, Civil Engineering Department, May 1991.

23. Peters, R.,; Shem L. Treatment of Soils Contaminated with Metals, Metal Speciation and Contamination of Soil Workshop, Jekyll Island Georgia, May 22-24, 1991

RECEIVED January 7, 1992

SURFACTANTS

Chapter 10

Factors Affecting Surfactant Performance in Groundwater Remediation Applications

Jeffrey H. Harwell

Institute for Applied Surfactant Research and School of Chemical Engineering and Materials Science, University of Oklahoma, Norman, OK 73019-0628

The effectiveness of a surfactant in enhancing the removal of a contaminant from the subsurface can be expected to be a function not only of the surfactant's interaction with the contaminants, but also of the surfactant's interaction with the aquifer media at the conditions in the aquifer. There are a number of types of surfactant behavior that can be anticipated as potentially reducing the surfactant's ability to remove the contaminant: precipitation, liquid crystal formation, formation of a coacervate phase, partitioning into trapped residual phases, or adsorption onto the aquifer's solid surfaces. These phenomena and the variables affecting them should be taken into account in the design of a surfactant system for any remediation project. Additionally, the interaction of contaminants with the surfactant solution formed in the aquifer should be anticipated to be highly dependent on temperature, electrolyte concentration, and the concentration of the contaminant.

Surfactants show significant potential for application in enhancing the remediation of contaminated ground water. Two very different technologies are being considered for achieving an enhancement: 1. Solubilization of contaminants in surfactant micelles and 2. Mobilization of residual liquids by reduction of the capillary forces trapping liquid droplets in the aquifer porous medium. The restrictions on potential surfactants are very different for the two technologies, yet both can be expected to show a strong dependence on the environment in an aquifer. When selecting a surfactant or a mixture of surfactants for a potential remediation project, one must account not only for the ability of the surfactant either to solubilize the contaminant or to produce ultra-low interfacial tensions with trapped phases, but one must also consider whether or not the surfactant will retain its activity in the aquifer at aquifer conditions. Besides the loss of surfactant activity from the adsorption of the surfactant on the solid surfaces in the aquifer, aqueous surfactant solutions can undergo a number of phase changes that will result in a significant loss of activity. The possible phase changes are precipitation of ionic surfactants by other ions, the formation of liquid crystals or a coacervate phase, and the abstraction of the surfactants from the aqueous phase into a

0097–6156/92/0491–0124$06.00/0

trapped phase. Unfortunately, some of the same changes in surfactant structure that increase solubilization power or lower interfacial tensions can also result in increases in surfactant losses by these mechanisms. Failure to take such possibilities into account in the design of a surfactant enhanced remediation project may result in the loss of the surfactant in the aquifer without the anticipated increase in the remediation rate.

Relationship between Solubilization and Surfactant Structure

Solubilization is a phenomenon associated with the formation of micelles. A micelle is a transient aggregate of 50 to 200 surfactant molecules. The concentration of surfactant at which micelles begin to form is called the critical micelle concentration (CMC); the CMC is a function of the structure of the surfactant, the temperature of the surfactant solution, the concentration of added electrolytes, and the concentration of solubilizates and other amphiphiles.

Partition Coefficients. The solubilizing power of micelles of a particular surfactant is often expressed in terms of a partition coefficient for the solubilizate between the micelles and the aqueous phase excluding the micelles. The partition coefficient is usually defined as the ratio of the mole fraction of solubilizate in the micelles (molecules of solubilizate per micelle/total solubilizate plus surfactant molecules per micelle) to the concentration of solubilizate in the solution outside of the micelles. This means the partition coefficient is not dimensionless, but has units of reciprocal concentration. Most of the literature on partition coefficients concerns the partition coefficient at only one solubilizate activity, the activity at which a separate solubilizate phase forms; i.e., at unit activity (1, 2). In actuality, the partition coefficient should be expected to be a function of the amount of solubilizate in the micelles (3). For non-polar solubilizates like cyclohexane, the partition coefficient tends to decrease with decreasing concentration of the solubilizate while for amphiphilic and polar solubilizates, the partition coefficient tends to increase with decreasing concentration of the solubilizate. (The fundamental reasons for these phenomena are explained below.)

Role of Solubilizate Type. When solubilization is to be the mechanism by which addition of a surfactant can be expected to enhance the remediation rate, almost any micelle-forming surfactant is a potential candidate for the application, in that the ability of a micelle to solubilize a particular compound is not greatly affected by the structure of the surfactants in the micelle. There are basically three locations within a micelle where contaminants may be solubilized: the hydrophobic core of the micelle, the polar surface of the micelle, and the so-called "palisade layer"--the transition region between the surface of the micelle and the core of the micelle (Figure 1). The location or combination of locations within the micelle at which a particular contaminant will be solubilized depends primarily on the water solubility of the contaminant molecules.

When the contaminant is non-polar and has low solubility in water, such as octane, then it may be expected to concentrate in the hydrophobic core of the aggregate. Since increasing the size of the core will increase the room available for solubilized molecules, an increase in the size of the hydrocarbon moiety of the surfactant can be expected to increase the moles of contaminant solubilized per mole of surfactant. Also, since unaggregated surfactant molecules (monomers) do not contribute significantly to solubilization, partition coefficients are ususally higher with surfactants which have lower CMC values. However, changes in the hydrophobic moiety can also be expected to effect the surfactant's tendency to be lost by adsorption, precipitation, coacervate formation, or partitioning into trapped nonaqueous phases. Such losses may more than offset any gains in solubilizing power and should be considered

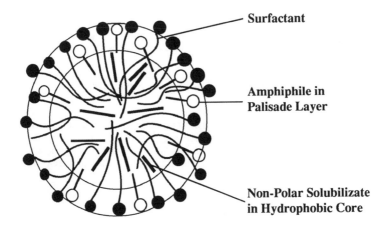

Figure 1.   Solubilization Loci

in conjunction with solubilization in making decisions about surfactant selection for a particular remediation effort. The partition coefficient for non-polar solubilizates tends to increase with solubilizate concentration because as more solubilizate is incorporated into the core of the micelle, the core becomes more like the solubilizate and the "solvent power" of the core for the solubilizate increases.

When the molecules to be solubilized are amphiphilic, such as hexanol, then the solubilized molecules can be expected to be solubilized in the palisade layer of the micelles. In this case the spaces between the headgroups of the surfactants serve as sites for the solubilizate molecules and the use of surfactant molecules with bulky headgroups may increase the sites for solubilization at the micelle's surface, thereby increasing the moles of solubilize molecules per mole of surfactant molecules. Again, however, if the change in headgroup increases the susceptibility of the surfactant molecule to one of the mechanisms by which surfactant may be lost in the aquifer, the losses may more than offset the advantages of the increase in solubilizing power.

The incorporation of amphiphilic solubilizates in the palisade layer also explains why the partition coefficient for amphiphiles decreases with increasing concentration of solubilizate: Solubilizate molecules compete with one another for the available sites in the palisade layer of the micelles.

In summary, then, since solubilization occurs because of very non-specific micelle/solubilizate interactions that are present whenever micelles are present, it is advisable to emphasize the phase behavior of the surfactant at the aquifer conditions and the interaction of the surfactant with the phases in the aquifer in selecting a surfactant for a remediation process.

Relationship Between Interfacial Tensions and Surfactant Structure

Surfactants can also be used to release NAPLS (Non-Aqueous Phase Liquids, such as gasoline) or DNAPLS (Dense Non-Aqueous Phase Liquids, such as tri-chloroethylene) trapped in the pores of an aquifer medium. The capillary forces that cause trapping of a non-aqueous liquid depend primarily on the interfacial tension between the trapped liquid and the aquifer water. Only when the interfacial tension is reduced to an ultra-low level (0.001 mN/m or less) can this mechanism be expected to dramatically improve the performance of a remediation project. It is now known that such ultra-low interfacial tensions are found only in the presence of a middle phase microemulsion (4-7). Microemulsion formation requires a dense surfactant film be adsorbed at the interface between the liquids. This will occur only when the hydrophilic and lipophilic groups of the surfactant are balanced so that the solubility of the surfactant in either of the two liquid phases is not too high (6). For a given hydrophilic group, this requirement greatly restricts the possible hydrophobic groups that will yield molecules which will give adequate lowering of the interfacial tensions. Systematic procedures for designing a surfactant system to give an ultralow interfacial tension (or, equivalently, to form a middle phase microemulsion) have been developed, especially since the decade of the seventies, and can be found in the literature; it is beyond the scope of this brief paper to review them (4, 5).

What must be emphasized is that--in contrast to the situation with solubilization--a surfactant that will not exhibit excessive losses through precipitation, phase trapping, etc., cannot in general be expected to also lower interfacial tensions in the aquifer to the point where trapped non-aqueous phases will be mobilized. Only when the surfactant structure and the design of the surfactant system carefully account for the specific NAPL or DNAPL (7), the composition of the ground water, and the temperature of the aquifer will ultralow interfacial tensions be obtained. When solubilization is the enhanced remediation mechanism, one can afford to focus on minimizing surfactant losses when optimizing the selection of a surfactant; when mobilization is the enhanced remediation mechanism, one must simultaneously consider obtaining ultralow interfacial tensions and minimizing surfactant losses.

Surfactant Loss Mechanisms: Adsorption on the Aquifer's Solid Surfaces

Because surfactants are surface active, their adsorption on the porous medium of the aquifer can be minimized, but cannot be eliminated. Three types of interactions are involved in the adsorption of a surfactant at at solid/liquid interface: the attractive or repulsive interaction between the hydrophilic group and the surface, the attractive interaction between the hydrophobic group and the surface, and the lateral interactions which occur between adsorbed surfactants (8). For surfactants, the most important interactions between the hydrophilic groups and the surface are electrostatic, hydrogen bonding, and complexation.

Electrostatic Interactions. Electrostatic interactions are most important for ionic surfactants. The magnitude of these interactions are strongly dependent on the nature of the aquifer medium (sandstone versus carbonate, for example), on the pH of the ground water, and on the presence of multivalent ions; it is easy to over generalize in this area (9-12). Most natural surfaces are negatively charged, however, under naturally occurring conditions (13). As a result, anionic surfactants will experience a repulsive electrostatic interaction with most natural surfaces; this serves to make them adsorb to a lesser extent than cationic or nonionic surfactants for most applications. In contrast, cationic surfactants experience an attractive electrostatic interaction and show high levels of adsorption when compared to anionic or nonionic surfactants of the same size hydrophobic moiety. This generalization correctly predicts the relative adsorption of anionics and cationics on SiO_2 surfaces such as are found in sandstone. On dolomite or calcite, however, at low to neutral pH values the solid/liquid interface is positively charged, and anionics will be higher adsorbing than similar cationics.

Hydrogen Bonding Interactions. The adsorption of nonionics depends greatly on the extent of interaction between hydrogen bonding groups on the solid surface and the aqueous solution. For alumina surfaces, which have high heats of immersion, water displaces alcohol groups and polyethyleneoxide groups from the interface, so that polyethyleneoxide nonionics are almost nonadsorbing on aluminas. On SiO_2 surfaces, which have lower heats of immersion, the nonionic groups will displace water at low concentrations of surfactant, so that nonionics will readily form complete bilayers of adsorbed surfactants on most silica surfaces. One should anticipate, therefore, that in most sandstone reservoirs nonionics would show a higher degree of adsorption than anionic surfactants.

Hydrophobic Interactions. Figure 2 is a schematic representation of a typical surfactant adsorption isotherm (9). It is commonly divided into four regions. Region 1 is a low adsorption density region in which the surfactant adsorption shows a linear dependence on surfactant concentration. This is indicative of an absence of lateral interactions between the adsorbed surfactant molecules. Region 2 is indicated by an increase in the slope of the adsorption isotherm; this increase in slope is indicative of the onset of attractive lateral interactions between the surfactants that results in the formation of micelle-like surfactant aggregates at the solid/liquid interface. These aggregates are called hemimicelles or admicelles, depending on their morphology (12). All of the factors that result in the lowering of the CMC of a surfactant can be expected to increase the formation of admicelles, also. Region 3 is a region where the slope of the isotherm decreases because of the onset of repulsive lateral interactions between admicelles. Region 4 is a region where the adsorption reaches a plateau either because the surface has become saturated with admicelles or because the surfactant concentration in the solution has reached the CMC of the surfactant. At the CMC the first micelle forms at the same chemical potential as the last admicelle to form; if the surfactant concentration is increased further, this only results in the formation of more micelles at the same chemical potential as that of the first micelle to form, without an increase in the adsorption of more surfactant.

While Figure 2 is a typical surfactant adsorption isotherm, only parts of it may be present in some situations. When the surfactant has a very low adsorption (as, for example, when there is electrostatic repulsion between the surfactant and the surface) the CMC may be reached before the onset of lateral interactions between adsorbed surfactants; in this case regions 2 and 3 will be missing from the isotherm. If the surfactant has a very high adsorption on the surface, the Region 1/Region 2 transition may occur below the concentration at which the surfactant can be detected, so that, while Region 1 may exist, adsorption at that level cannot be measured.

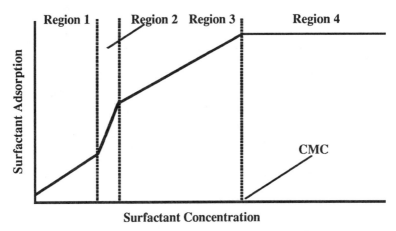

Figure 2. Schematic of a Typical Surfactant Adsorption Isotherm.

Minimizing Adsorption. In selecting a surfactant to minimize losses by adsorption, the single most important factor is the nature of the hydrophilic group. If the surface is charged (and it will usually be charged), then the headgroup of the surfactant should be like-charged to the surface to minimize the adsorption. Beyond the selection of the headgroup, factors that increase the CMC also decrease the adsorption: decrease in electrolyte concentration, decrease in the concentration of oppositely charged multivalent ions, decrease in the size of the hydrophobic moiety of the surfactant, an increase in the amount of branching in the hydrophobic moiety. Unfortunately, most of these changes will also result in a decrease in the solubilizing power of the surfactant. Similar considerations will affect the adsorption of nonionic surfactants, except that the electrostatic interactions become relatively unimportant.

Surfactant Loss Mechanisms: Precipitation

While the use of an ionic surfactant can minimize losses by adsorption, it introduces the problem of losses by precipitation (14-19). Anionic surfactants are especially susceptible to precipitation by multivalent cations like Ca^{++} and Mg^{++}. The precipitation of a single surfactant species is relatively easy to understand below the CMC of the surfactant. Above the CMC the cations may associate with the micelles of the surfactant to such an extent that the activity of the cations is reduced to the point that precipitate no longer forms. When multiple surfactant species are present, the formation of mixed micelles can lead to an unexpectedly high tolerance for multivalent cations.

Factors Affecting Precipitation. As with adsorption, there are a variety of changes in the surfactant that can be introduced to minimize the tendency of the molecule to precipitate. Many of these changes are similar in nature to those needed to minimize adsorption: decrease in electrolyte concentration, decrease in the concentration of oppositely charged multivalent ions, decrease in the size of the hydrophobic moiety of the surfactant, increase in the amount of branching in the hydrophobic moiety. Surprisingly, however, an increase in the concentration of monovalent counterions can also decrease the tendency to precipitate. This is because the monovalent ions can lower the CMC of the surfactant without causing precipitation (up to a point, of course). If the CMC is reduced enough, the monomer concentration can drop below that required to exceed the Ksp of the monomer/multivalent counterion pair. The use of surfactant mixtures--combining, for example, an anionic and a nonionic surfactant-- can also have dramatic effects on the monomer concentration of the anionic surfactant, which greatly reduces its susceptibility to precipitate formation.

Krafft Points. An important parameter to evaluate in considering the use of a surfactant in a particular application is the Krafft Point of the surfactant. The Krafft Point is the temperature at which the solubility of the monomer in water reaches the CMC of the surfactant. At temperatures below the Krafft Point the surfactant will show very limited solubility, while above the Krafft Point, the solubility will be much greater. The closer the temperature of a solution is to the Krafft Point of a surfactant, the greater will be the surfactant's tendency to precipitate.

Co-Surfactants. Another approach to reducing precipitation is to add a so-called co-surfactant to the surfactant system. A co-surfactant is a low molecular weight, amphiphilic molecule with high water solubility like sec-butanol or iso-propanol. Molecules like this fit into the palisade layer of micelles with the result that they lower the CMC of the surfactant; since they are excluded from the tightly packed headgroup region of precipitate, they do not lower the free energy of the precipitate phase. Additionally, the hydrophobic moiety of the co-solvent increases the solubility of the hydrophobic moiety of the surfactant monomer, thereby raising the Ksp. By all of these mechanisms a co-solvent decreases the tendency of ionic surfactant to be lost by precipitate formation. Unfortunately, if micelles are needed for their solubilizing power, the co-surfactant will compete with amphiphilic solubilizates for the palisade layer sites.

Surfactant Loss Mechanisms: Liquid Crystal Formation

Surfactants also have a tendency to separate from an aqueous solution by forming a thermodynamically stable phase called a "liquid crystalline phase" (20, 21). The principle types of surfactant liquid crystalline phases may be thought of as concentrated solutions of spherical, rod-like, or disk-like micelles; these are the viscous-isotropic phase (composed of spherical micelles in a body-centered, close-packed arrangement), the hexagonal phase (composed of hexagonally close-packed cylindrical micelles) and the lamellar or neat phase (composed of sheets of surfactants in bilayers with water and disassociated ions between the layers). These phases can exist in equilibrium with dilute aqueous surfactant solutions. They are in general very viscous and have a density greater than that of water. When only water and surfactant are present, they tend to form only at relatively high surfactant concentrations and at temperatures above the Krafft point of the salt of the ionic surfactant involved. When solubilizate is present, however, they form more readily. The type of solubilizate present, the structure of the surfactant molecule, the temperature, and the type and concentration of electrolytes present all play a role in determining whether a liquid crystalline phase will form. In general, however, conditions under which surfactant precipitation is likely become conditions favorable to liquid crystal formation when high concentrations of solubilizate are present.

Surfactant Loss Mechanisms: Coacervation

The formation of a coacervate phase is most commonly encountered with nonionic surfactants. When the solubility of the monomers of a nonionic surfactant drops below the CMC of the surfactant, the nonionic surfactant will phase separate from the aqueous solution in another aqueous surfactant phase. This separate aqueous surfactant phase may contain on the order of 90 weight percent surfactant, may be highly viscous, and may be more dense than water. Formation of a coacervate phase corresponds to a large drop in the activity of the surfactant. The tendency of a nonionic surfactant to form a coacervate phase is related to its cloud point temperature: When an aqueous solution of a nonionic surfactant is heated, there may be a high temperature at which, if the solution is heated to that point, the solution becomes cloudy. This temperature is the cloud point temperature. If allowed to stand at or above the cloud point temperature, a separate coacervate phase may form. In the presence of a low solubility solubilizate, the cloud point temperature of a solution of nonionic surfactant will decrease; ie, the tendency of the surfactant to form a separate phase will increase. If electrolyte is added to the solution, this will also decrease the cloud point: the cloud point temperature depends on the concentration of electrolyte (22).

The cloud point temperature is also called the lower consolute solution temperature. Though not generally reported as part of the characterization of nonionic surfactants, there may also be a low temperature at which, if the surfactant solution is cooled to that point, the solubility of the monomer drops below the CMC of the surfactant and the surfactant will again form a separate phase; this is an upper consolute solution temperature. In contrast to the cloud point temperature, the upper consolute solution temperature will be raised by changes in the solution that lower the solubility of the surfactant monomer.

When selecting a nonionic surfactant for application in a ground water remediation project, one must be careful to take into account the phase behavior of the surfactant at aquifer conditions. Unfortunately, solubilization parameters tend to become larger as the temperature of a solution of nonionic surfactant nears a cloud point temperature; selecting a nonionic surfactant for a project from a homologous series of nonionic surfactants on the basis of the surfactant which has the highest solubility parameter may be equivalent to choosing the system with the greatest tendency to form a coacervate phase.

Surfactant Loss Mechanisms: Partitioning into the Trapped Phase

In the design of surfactant enhanced remediation project, the solubility of the surfactant in the nonaqueous phase should also be considered. Though ionic surfactants will generally have only a very low solubility in most nonaqueous liquids, nonionic surfactants may have high solubilities in nonaqueous liquids (6). The partition coefficient for the surfactant between the aquifer's aqueous phase and the NAPL or DNAPL should also be considered when selecting a surfactant. In the presence of sufficient amount of the nonaqueous phase, the trapped nonaqueous phase may extract the surfactant from the groundwater and retain it. As a rule of thumb, nonionic surfactants with high water solubilities will have both high cloud point temperatures and smaller partition coefficients for partitioning into nonaqueous phases. Ionic surfactants are not entirely immune to this loss mechanism. The divalent metal salts of ionic surfactants can have significant solubilities in nonaqueous liquids; divalent salts of the fatty acid surfactants are added to motor oils for corrosion inhibition.

Summary

Failure to take into account the possible mechanisms by which surfactants can loose their activity in an aquifer environment can result in the choice of a surfactant for a remediation project that will fail to perform as expected in the application. It is critical that the phase behavior of the surfactant at aquifer conditions be understood. Avoiding surfactants with unfavorable phase behavior is more easily accomplished when solubilization is the enhanced remediation mechanism, because solubilization does not require a close matching of the structure of the surfactant and the structure of the solubilizate, leaving room for selection of a surfactant structure with favorable phase behavior. This is a much more difficult problem when formation of a microemulsion phase is necessary for mobilization of residual nonaqueous liquids as part of a remediation project, as the structure of the surfactant must be balanced to give form a microemulsion with the particular nonaqueous liquid and the aquifer aqueous medium at aquifer conditions.

Acknowledgments

The author wishes to thank the corporate sponsors of the Institute for Applied Surfactant Research for their financial support, which made this paper possible: DuPont, Sandoz, Union Carbide, and Kerr-McGee Corporations.

Literature Cited

1. *Micellization, Solubilization, and Microemulsions*; Mittal, K. L., Ed.; Plenum: New York, NY, 1977; Vols. 1 and 2.
2. *Solution Behavior of Surfactants*; Mittal, K. L., Fendler, E. J., Eds.; Plenum: New York, NY, 1982; Vols. 1 and 2.
3. Nguyen, C. M.; Christian, S. D.; Scamehorn, J. F. *Tensides, Surfactants, Deterg.*, **1988**, *25*, 328.
4. Bourrel, M.; Chambu, C. *Soc. Petr. Engin. J.*, **1983**, *23*, 327.
5. Puerto, M. C.; Reed, R.L. *Soc. Petr. Engin. J.*, **1983**, *23*, 669.
6. Graciaa, A.; Lachaise, J.; Marion, G.; Schechter, R.S. *Langmuir*, **1989**, *5*, 1315.

7.   Bourrel, M.; Verzaro, F.; Chambu, C. *SPE Reservoir Engin.*, **1987**, *2*, 41.
8.   Aveyard, R. In *Solid/Liquid Dispersions*, Tadros, T. F., Ed.; Academic Press: London, UK, 1987, Ch. 5.
9.   Scamehorn, J. F.; Schechter, R. S.; Wade, W. H. *J. Colloid Interface Sci.*, **1982**, *85*, 463.
10. Scamehorn, J. F.; Schechter, R. S.; Wade, W. H. *J. Colloid Interface Sci.*, **1982**, *85*, 479.
11. Scamehorn, J. F.; Schechter, R. S.; Wade, W. H. *J. Colloid Interface Sci.*, **1982**, *85*, 494.
12. Bitting, D.; Harwell, J. H. *Langmuir*, **1987**, *3*, 500.
13. Rosen, M. J. *Surfactants and Interfacial Phenomena, Second Edition*, John Wiley and Sons: New York, NY, 1989, Ch. 2.
14. Scamehorn, J. F.; Harwell, J. H. In *Mixed Surfactant Systems*; Ogino, K.; Abe, M., Eds.; Marcel Dekker: New York, NY, 1991, "Surfactant Precipitation".
15. Scamehorn, J. F. In *Phenomena in Mixed Surfactant Systems*; Scamehorn, J. F., Ed.; American Chemical Society: Washington, DC, 1986, Ch. 1.
16. Peacock, J. M.; Matijevic, E. *J. Colloid Interface Sci.*, **1980**, *77*, 548.
17. Noik, C.; Baviere, M.; Defives, D. *J. Colloid Interface Sci.*, **1987**, *115*, 36.
18. Matheson, M. L.; Cox, M. F.; Smith, D. L. *J. Amer. Oil Chem. Soc.*, **1985**, *62*, 1391.
19. Stellner, K. L.; Amante, J. C.; Scamehorn, J. F.; Harwell, J. H. *J. Colloid Interface Sci.*, **1988**, *123*, 186.
20. *Proceedings of the International School of Physics (1983: Varenna, Italy)*, Degiorgio, V., Ed.; Elsevier Science Publishers: Amsterdam, The Netherlands, 1985, "Physics of Amphiphiles–Micelles, Vesicles and Microemulsions".
21. Benton, W. J.; Faijal, S. K.; Ghosh, O.; Qutubuddin, S.; Miller, C. A. *SPE Reservoir Engin.*, **1987**, *2*, 664.
22. Gullickson, N. D.; Scamehorn, J. F.; Harwell, J. H. In *Surfactant-Based Separation Processes*; Scamehorn, J. F.; Harwell, J. H., Eds.; Marcel Dekker: New York, NY, Ch. 6, 1989.

RECEIVED January 24, 1992

Chapter 11

Influence of Surfactant Sorption on Capillary Pressure—Saturation Relationships

F. N. Desai, A. H. Demond, and K. F. Hayes

Department of Civil and Environmental Engineering, University of Michigan, Ann Arbor, MI 48109–2125

The capillary pressure-saturation relationship, a fundamental relationship in the description of multiphase flow, depends on the interfacial properties of the system. Sorption of a cationic surfactant such as cetyltrimethylammonium bromide (CTAB) at the various interfaces of a system changes interfacial properties such as electrophoretic mobility, interfacial tensions, and contact angle. The objective of this paper is to examine the effect of the changes in these interfacial properties on the capillary pressure-saturation relationships for the air-water-silica system. The results presented here show that as the sorption of CTAB increases, the naturally negatively-charged silica surface becomes positively charged. This change in charge is reflected in the contact angle which passes through a maximum when the electrophoretic mobility is close to zero. The spontaneous imbibition capillary pressure relationship is more sensitive to changes in interfacial properties than the drainage relationship. In the air-water-silica system studied here, no imbibition is observed at the maximum contact angle. The surface tension and contact angle can be used to predict both the drainage and imbibition relationships of the air-water-silica-CTAB systems from that of the air-water-silica system. The prediction is accomplished through scaling using the value of surface tension and the operational contact angle, which can be obtained from the intrinsic angle through the incorporation of corrections for roughness and interfacial curvature. A comparison of the measured and calculated capillary pressure relationships shows that it is possible to predict the effect of surfactant sorption on both drainage and imbibition capillary pressure-saturation relationships for the system studied.

To describe the movement of water in the vadose zone, Richards' equation is commonly used. To apply this equation, additional information in the form of constitutive relationships must be provided. Much research has been devoted to the description and prediction of the constitutive relationship of capillary pressure-saturation (1-3). However, many of these efforts seem implicitly to regard the soil matrix as strongly water-wet. But that state, in reality, depends on the nature of the soil solids, the composition of the subsurface water and the properties of solutes

0097–6156/92/0491–0133$06.00/0

present. If a surface-active compound is added to or naturally occurs in the subsurface, it will sorb differentially at the three interfaces of the system: air-water, water-solid and air-solid. The exact degree of sorption will be influenced by the pH and ionic strength of the aqueous solution. The sorption will, in turn, alter the interfacial tensions and the contact angle of the system, consequently impacting the capillary pressure-saturation relationships.

Since the interfacial tension between a fluid and a solid cannot be measured directly, surrogate measurements, such as electrophoretic mobility or zeta potential, are commonly used as indicators of interfacial conditions. A number of studies have related sorption or zeta potential to the magnitude of the contact angle in air-water-mineral-surfactant systems (4-7). These studies showed that the sorption of cationic surfactants to a negatively-charged mineral surface caused the surface charge to become positive. The increase in the surface charge resulted in the mineral surface's becoming first less water-wet and then strongly water-wet once again. Similar effects are also noted in organic liquid-water-mineral studies, but the range of wettability changes which result from the adsorption of surface-active compounds may be even greater (8-9). While in each case cited above the effects of sorption on wettability were demonstrated, the implications of these changes with respect to capillary pressure were not explored.

On the other hand, Morrow (10-11) and Demond and Roberts (12) have looked at the effects of interfacial tension and contact angle on capillary pressure-saturation relationships. Their work showed that the contact angle affects the spontaneous imbibition relationship more strongly than the primary drainage. Above a contact angle of about 65°, no spontaneous imbibition occurred. However, in these studies, the changes in contact angle and interfacial tension were accomplished through the use of different organic liquids, rather than through the sorption of a surface-active compound. Furthermore, Demond and Roberts (12) demonstrated a means of scaling capillary pressure-saturation relationships using the liquid-liquid interfacial tension and contact angle. Again, this methodology was developed for organic liquid-water systems and its applicability to a system where the wettability changes are governed by sorption reactions has yet to be demonstrated.

Thus, previous studies have focused on the relationship between sorption and contact angle, or on the relationship between contact angle and capillary pressure. The purpose of this paper is to integrate these foci and examine the effects of surfactant sorption on capillary pressure by correlating measurements of sorption, electrophoretic mobility, surface tension and contact angle, with changes in the capillary pressure-saturation relationship for the air-water-silica system. Because sorption has a direct effect on wettability and wettability impacts capillary pressure relationships, it should be possible to develop a quantitative relationship among sorption, wettability, and capillary pressure.

Background

The presence at waste disposal sites of a variety of surface-active species, such as organic acids, chelating agents, and humic substances, is not uncommon (13). Despite this observation, little work has been devoted to understanding the effects of sorption of these types of solutes on the movement of water in the vadose zone. Progress has been made in the numerical solution of Richards' equation (14-15); yet, little attention has been devoted to the changes in key constitutive properties such as the capillary pressure-saturation and relative permeability-saturation relationships that might result from the presence of surface-active contaminants. Some recent studies examine the changes in these relationships which occur due to the introduction of a separate organic liquid phase (12, 16-18). Nevertheless, this work is still primarily oriented towards developing models for these constitutive properties, rather than towards understanding the phenomena governing the behavior of these relationships. Ultimately, to understand the behavior of these macroscopic relationships, it is necessary to examine the microscale parameters influencing them.

Richards' equation may be written:

$$\frac{\partial \theta}{\partial t} - \nabla \cdot K \nabla P_W - \frac{\partial K}{\partial z} = 0 \tag{1}$$

where $\quad K \quad =\quad$ unsaturated hydraulic conductivity, which may be expressed as relative permeability, k_r, where $k_r = K/K_{sat}$, and $K_{sat} =$ saturated hydraulic conductivity,

$\quad P \quad = \quad$ pressure head,

$\quad \theta \quad = \quad$ moisture content of wetting phase, often expressed in terms of saturation, S, where $S = \theta/n$, and n = porosity, and subscript

$\quad W \quad = \quad$ wetting phase, usually water.

To solve this equation, additional information must be provided in the form of constitutive relationships, such as the capillary pressure-saturation relationship:

$$P_{NW} - P_W = P_C = f(S) \tag{2}$$

where the subscripts

$\quad c \quad = \quad$ capillary, and

$\quad NW \quad = \quad$ nonwetting.

in which capillary pressure, defined as the difference between the pressures in the wetting and nonwetting fluids, averaged over a representative elementary volume, is expressed as a function of saturation.

Although this relationship is macroscopic, based on a continuum approach to describing flow in porous media, it represents the effect of pore-level interfacial properties. At the scale of a single pore, the capillary pressure is described by the Young-Laplace equation:

$$P_C = \gamma_{H_2O/AIR} (1/R_1 + 1/R_2) \tag{3}$$

where $\quad R \quad = \quad$ principal radius of curvature,

$\quad \gamma \quad = \quad$ interfacial tension,

subscripts $\quad AIR \quad = \quad$ air phase, and

$\quad H_2O \quad = \quad$ aqueous phase.

If the meniscus is spherical and the pore cylindrical, then equation 3 may be rewritten as:

$$P_C = \frac{2\,\gamma_{H_2O/AIR}\,\cos\theta}{r} \tag{4}$$

where $\quad \theta \quad = \quad$ contact angle, and

$\quad r \quad = \quad$ radius of pore.

The capillary pressure across a curved meniscus depends upon the contact angle which, according to Young's equation, is a function of the interfacial tensions of the system:

$$\gamma_{H_2O/AIR} \cos\theta = \gamma_{AIR/S} - \gamma_{H_2O/S} \tag{5}$$

where the subscript

$$S \quad = \quad \text{solid.}$$

In turn, these interfacial tensions are affected by the sorption density at the three interfaces of the system according to the Gibbs adsorption equation:

$$d\gamma = -RT \left(\sum_i \Gamma_i \, d(\ln a_i) \right) \tag{6}$$

where
$$
\begin{aligned}
a_i &= \text{activity of species, i, in solution,} \\
R &= \text{gas constant,} \\
T &= \text{temperature, and} \\
\Gamma_i &= \text{sorption density of species, i, at a given interface.}
\end{aligned}
$$

Since the sorption density will typically increase with increasing activity or concentration of the surface-active compound, sorption at interfaces will usually result in the lowering the interfacial tension according to equation 6.

Equations 1-6 delineate a causal relationship between the advective movement of water in the vadose zone and sorption. Thus, to understand how the presence of surface-active compounds might influence the movement of water in the vadose zone implies examining the effect of sorption and the subsequent interfacial tension reduction on constitutive relationships such as capillary pressure-saturation.

The equations describing interfacial properties at a microscale cannot be applied directly to the macroscopic description of flow in porous media. Accurate values of the interfacial tensions of solids are difficult to obtain and surrogate measurements of surface conditions, such as electrophoretic mobility, are usually required. The contact angle defined by Young's equation for a smooth flat plate is often not the operative contact angle in a porous medium. For example, the roughness of a natural surface causes the contact angle to be hysteretic, showing an enlarged angle if advancing and a diminished angle if receding. In addition, the magnitude of the contact angle necessary to make equation 4 an equality is affected by the configuration of the pore. These factors make direct quantitative computations using the above equations difficult.

Yet with this framework in mind, methods have been developed for relating operative contact angles to intrinsic contact angles. Recent measurements of contact angles on rough surfaces and in bead packs have demonstrated that the operative contact angles show a systematic dependence on the Young's, or intrinsic, contact angle (19-22). Extending this work, Demond and Roberts (12) developed a methodology for calculating operative contact angles for drainage and imbibition in sands. They combined the curvature correction factor, $Z(\theta)$, defined by Melrose (20) to express the effect that the contact angle has on drainage or imbibition in terms of deviations from the capillary tube model:

$$Z(\theta) \equiv \frac{P_C\, r}{2\gamma_{H_2O/AIR}\cos\theta} \tag{7}$$

where   Z   =   curvature correction factor, where $Z \geq 1.0$
                        for drainage, and $0 \leq Z \leq 1.0$ for imbibition,

with roughness corrections determined empirically by Morrow (*21*) (Table I):

Table I. Roughness Corrections for Well-Roughened Surfaces

Intrinsic Angle	Operational Angle	
	drainage (receding angle)	imbibition (advancing angle)
$0° < \theta_{INT} < 21.6°$	$\theta_r = 0°$	$\theta_r = 0°$
$21.6° < \theta_{INT} < 87.6°$	$\theta_r = 0.5\exp(0.05\theta_{INT}) - 1.5$	$\theta_r = 2(\theta_{INT} - 21.6)$

where subscripts
                        INT   =   intrinsic, and
                        r   =   corrected for roughness.

to produce a means of calculating the operational angles for both drainage and imbibition. The operational contact angles can then be combined with a form of Leverett's (*23*) equation, written here assuming that the porosity and permeability of the systems are equal,

$$P_c(S_e)_2 = P_c(S_e)_1\, (\gamma_2/\gamma_1) \tag{8}$$

where S_e =   effective saturation = $(S - S_r)/(1 - S_r)$,
                        S_r =   residual saturation of the wetting phase,

to yield:

$$P_c(S_e)_2 = P_c(S_e)_1 \frac{[\gamma\cos\theta_r\, Z(\theta)]_2}{[\gamma\cos\theta_r\, Z(\theta)]_1} \tag{9}$$

Equation 9 provides a means of calculating the capillary pressure of one fluid-fluid pair in a porous medium from the capillary pressure of another fluid-fluid pair in the same porous medium. This methodology has proven useful in organic liquid-water systems, but has yet to be applied to systems which include surface-active compounds. If equation 9 is applicable to air-water-silica-surfactant systems, then the changes in capillary pressure resulting from the sorption of surfactants can be estimated based on a knowledge of the relationship between sorption and $\gamma_{H_2O/AIR}$ and θ.

Experimental Materials and Methods

Materials. The solid quartz (SiO_2) was selected as the mineral for this study because it represents one of the most commonly occurring minerals in groundwater environments, and has surface properties representative of many aluminosilicates. The surface of quartz is composed of silanol surface functional groups which are deprotonated at all pH values normally encountered in subsurface environments, thus imparting a negative charge to the quartz surface. In addition, quartz is available in many forms, such as smooth flat plates necessary for contact angle measurements and different particle sizes necessary for capillary pressure and sorption density measurements. Cetyltrimethylammonium bromide (CTAB)

$(CH_3(CH_2)_{15}N^+(CH_3)_3Br^-)$ was selected as the surfactant for these studies, for complementary reasons. CTAB is a cationic surfactant which sorbs relatively strongly to a negatively charged surface. Consequently, it has the potential for causing well-defined changes in wettability.

Several forms of quartz were purchased and prepared for these studies. Two size fractions of particles, #40 and F-65 (U.S. Silica, Berkeley Springs, WV), were used. The #40 has a mean particle diameter of 7.3 μm and a surface area of 0.31 m^2/gm, and was used in the sorption and electrophoretic mobility studies. The F-65 has a particle size range of 106 to 425 μm and was used in the capillary pressure-saturation experiments. The particles were cleaned by washing several times with 0.1 N HCl followed by several washes with 15% hydrogen peroxide at a pH of about 4, to remove metal ions and organic matter (24). Smooth quartz slides were used for the contact angle measurements (Quartz Scientific, Fairport Harbor, OH); their cleaning is described in the section on contact angle measurements.

The purity of the CTAB, as purchased, was greater than 99% (Fluka Chemical, Hauppage, NY). Based on electrical conductivity measurements, the critical micelle concentration (CMC) of the CTAB was 9.5 x 10^{-4} M, which is in close agreement with the literature values (25). The CTAB was further purified by recrystallizing three times from a 4:1 mixture by volume of acetone and methanol (26). The radiolabeled CTAB was obtained from American Radiolabeled Chemicals (St. Louis, MO) and was used as purchased. All solvents used for cleaning were HPLC grade (Mallinckrodt Specialty Chemicals, Paris, KY). Ultrapure water was obtained by passing deionized, distilled water through Milli-Q (Millipore, Bedford, MA) cartridge filters.

Methods. To examine the relationship between sorption and capillary pressure, five types of measurements were made: sorption, electrophoretic mobility, surface tension, contact angle and capillary pressure-saturation. For the purposes of describing the experimental effort, these measurements are divided into three parts: 1) surface chemical measurements, 2) surface tension and contact angle measurements, and 3) capillary pressure-saturation measurements. Sodium chloride at a concentration of 0.01 M was used as the background electrolyte in all the experiments. The experiments were conducted at room temperature (23±2°C).

Surface Chemical Measurements. Adsorption isotherms were measured for CTAB on silica over a range of surface coverages at pH values from 5 to 9. These isotherms were obtained by mixing appropriate amounts of silica, aqueous solution containing radiolabeled CTAB, background electrolyte, and acid or base in 12 ml polypropylene centrifuge tubes, and equilibrating for over 24 hours. The slurry was centrifuged and the adsorption density calculated based on the fraction of the total radioisotope remaining in the supernatant. Measurements aimed at assessing the significance of CTAB losses to the sample container walls showed that these losses were negligible for aqueous phase surfactant concentrations in excess of 10^{-6} M.

Potentiometric titrations were performed to give the electrophoretic mobility as a function of pH for different total CTAB concentrations. These measurements were made using a Matec ESA-8000 (Matec Applied Sciences, Hopkinton, MA). By applying an alternating electrical field to a liquid, charged particles suspended in the liquid will oscillate, creating an alternating pressure wave. The signal created by such movement can be converted to electrophoretic mobility or zeta potential using the methodology developed by O'Brien (27).

At a particular total CTAB concentration, the silica was mixed with appropriate amounts of water, CTAB, NaCl, and acid to give a pH close to 5.5. The titration curve was obtained by incrementally adding NaOH in 0.1 pH unit aliquots and mixing for 5 mins (a time period which was sufficient to stabilize the ESA signal) after each addition. The process was repeated up to a pH of 9.5. This data was then combined with the adsorption data to give electrophoretic mobility as a function of aqueous CTAB concentration.

Surface Tension and Contact Angle Measurements. The surface tension and the contact angle on quartz of aqueous solutions of CTAB were measured for a range of CTAB concentrations at a pH of 6 and NaCl concentration of 0.01 M. These measurements were made using an automated pendant/sessile drop instrument based on the Axisymmetric Drop Shape Analysis (ADSA) technique (28). A droplet was formed with a microsyringe (Gilmont, Great Neck, NY). The droplet was then enlarged with a M3Z Plan S microscope (Wild Leitz, Heerbrugg, Switzerland). The enlarged droplet was photographed using a solid-state CCD monochrome camera (4810 Series, Cohu, San Diego, CA), and the image sent to a digitizer where the analog signal was converted to a digital signal containing the image data in the form of pixels. A microcomputer with a 80386 microprocessor acquired these data and performed the image analysis and computation. The computational analysis involved fitting Laplace's equation to an arbitrary array of coordinate points selected from the drop profile. An objective function that expresses the deviation of the physically observed curve from the Laplacian curve was minimized to give the best value for the surface tension or contact angle (29). To damp vibrations, the apparatus was placed on a vibration isolation table (Technical Manufacturing, Peabody, MA).

Since the cleaning procedures influenced the reproducibility of the results, considerable attention was devoted to developing rigorous techniques. To clean the quartz slides, syringe barrels, quartz cell and other glassware, they were soaked for 20 minutes each in methanol, acetone, and Milli-Q water, followed by overnight soaking in Chromerge solution (VWR Scientific, Chicago, IL). These items were then again thoroughly rinsed in Milli-Q water. The syringe barrels and other glassware were oven dried at 110°C. The quartz slides were not dried, but were instead stored in Milli-Q water. The stainless steel needles and teflon syringe plungers and stage were soaked for 20 minutes each in hexane, methanol, acetone, and Milli-Q water, after which the teflon parts were soaked in Micro (International Products, Trenton, NJ). After rinsing thoroughly with Milli-Q water, the teflon and steel parts were dried in the oven at 110°C.

The surface tension measurements were made by forming pendant drops of CTAB solution in air. These droplets were made using a teflon tube attached to a microsyringe adjusted with a micromanipulator (World Precision, New Haven, CT) within an environmental chamber (Rame-Hart, Mountain Lakes, NJ) which had been flushed with high purity nitrogen saturated with Milli-Q water. The surface tension dropped sharply in the first 5 minutes after the pendant drop was formed. After about 10 minutes, the surface tension became fairly constant, dropping by less than 0.3 dynes/cm every 10 minutes. Based on this, all the surface tension measurements were made at 10 minutes equilibration time.

For contact angle measurements, the quartz slide was removed from the Milli-Q water, secured to a teflon stage and placed in a quartz cell containing CTAB solution. The slide was allowed to equilibrate for 20 minutes with the CTAB solution. The needle attached to the microsyringe was passed through a teflon septum mounted on the stage and a sessile air droplet was formed on the underside of the quartz slide through a hole in the center of the slide. Both advancing and receding contact angles were measured. The measurements were made 10 minutes after the droplet was advanced or receded. After this time, the contact angle changed by less than one degree in 15 minutes.

Capillary Pressure-Saturation Measurements. The capillary pressure-saturation measurements were made using custom-made teflon-coated stainless steel pressure cells that were patterned after Tempe cells (*30*). The coreholder had a height of 3.0 cm and an interior diameter of 5.5 cm. It was consistently packed with cleaned, dry F-65 silica to a consistent porosity of 0.34, in contact with a 1-bar porous ceramic plate. The cell was evacuated and then an aqueous solution (pH = 6, NaCl = 0.01 M) containing CTAB at a particular concentration was pumped through the cell at a rate of 5 ml/min until influent and effluent CTAB concentrations were the same. The CTAB concentration was measured to an accuracy of 5% by two-phase titration with mixed indicator using sodium dodecyl sulfate as the titrant (*31*). To ensure that the system was at equilibrium, several additional pore volumes were pumped through the cell after this point.

The pressure in the aqueous phase in the cell was maintained constant, close to atmospheric pressure. The pressure in the air phase was controlled by compressed air, passed through particulate and hydrocarbon traps, regulated by two regulators in series (Model 23, Scott Specialty Gases, Plumsteadville, PA; Model 40-7, Moore Products, Spring House, PA), and monitored by a low pressure gauge (Series 66-050, Wallace and Tiernan, Belleville, NJ). The drainage relationship was measured by increasing the air pressure in small increments. After each increment in air pressure, the cell was allowed to equilibrate for at least 24 hours. The decrease in saturation within the porous medium was calculated based on the increase in weight of the bottle collecting the effluent. This procedure was repeated until a negligible efflux was achieved in response to a step increment in pressure, suggesting that residual saturation had been achieved. This point served as the beginning of spontaneous imbibition. After each small reduction in air pressure, the system was allowed to equilibrate and the increase in saturation was calculated based on the weight loss of the effluent bottle.

Results and Discussion

Surface Chemical Measurements. Figures 1 and 2 give the adsorption density and electrophoretic mobility as a function of aqueous CTAB concentration over the pH range of 6 to 9. Silica is negatively charged above its point of zero charge, which is in the vicinity of pH 2 (*32*). Therefore, above pH 2, a cationic surfactant such as CTAB will show strong affinity for silica, giving an isotherm shape that has been previously noted as characteristic of adsorption in these systems (*33-34*). The adsorption mechanism can be inferred from the adsorption isotherm and the electrophoretic mobility data. The adsorption isotherm can be divided into four regions (Figure 3). In region I, the adsorption is characterized by electrostatic attraction between the negatively charged surface and the positively charged head group of isolated CTAB molecules. As the negative values of the electrophoretic mobility measurements suggest, the surface does not yet have a monolayer coverage. In region II, the marked increase in adsorption is mainly due to hemimicelle formation, a result of lateral interaction between the surfactant's hydrocarbon tails. Also in this region, the electrophoretic mobility becomes increasing less negative and finally turns positive, with the isoelectric point occurring at a CTAB concentration of about 1.5×10^{-4} M. In region III, the slope in the isotherm decreases suggesting a lessening tendency for a positively charged surfactant to adsorb onto a increasingly positively charged surface. The adsorption that does occur in this region results primarily from the hydrophobic interaction between the hydrocarbon tails of the surfactant molecules. This strong hydrophobic interaction leads to formation of bilayer clusters of surfactant molecules on the surface. Finally, the adsorption isotherm reaches a plateau in region IV at an

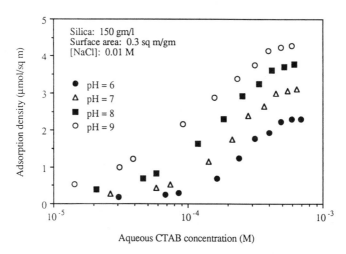

Figure 1. Adsorption of CTAB on silica as function of pH

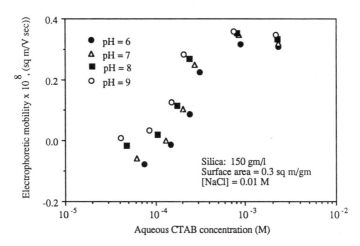

Figure 2. Electrophoretic mobility of CTAB-coated silica as function of pH

aqueous solution concentration of 5×10^{-4} M, when the maximum number of bilayer clusters is formed on the surface. Here, the adsorption density on the silica surface remains constant because additional molecules of surfactant introduced into the solution phase form solution-phase micelles rather than adsorbing to the solid surface.

Surface Tension and Contact Angle Measurements. Figure 4 shows the surface tension of aqueous solutions of CTAB at pH 6. The functional dependence shown in this figure is characteristic of aqueous solutions of surfactants (*35*). The surface tension decreases log-linearly from a CTAB concentration of about 1×10^{-5} M to 2.2×10^{-4} M, after which the surface tension is almost constant. These data suggest that the critical micelle concentration (CMC) of CTAB at pH 6 and NaCl = 0.01 M is 2.2×10^{-4} M, which compares favorably with the value of 2.7×10^{-4} M reported by McCaffery and Cram (*36*).

Figure 5 shows the advancing and receding contact angles of water on quartz as a function of CTAB concentration at pH 6. This figure indicates that, at low CTAB concentrations of about 1×10^{-6} M, the silica surface is highly hydrophilic. A comparison of Figures 2 and 5 shows that this portion corresponds to the region where the silica is negatively charged. As the CTAB concentration increases, the contact angle increases. At an aqueous phase CTAB concentration of about 1×10^{-4} M, which is close to the isoelectric point, the advancing contact angle passes through a maximum of 64°. At concentrations close to the CMC, the contact angle decreases, suggesting that the formation of a surfactant bilayer makes the surface strongly hydrophilic again.

These observations are not unilaterally corroborated by previous investigators. For a system of air-water-quartz-CTAB (in the absence of any background electrolyte), McCaffery and Mungan (*37*) observed a maximum advancing contact angle of 60°, which occurred at a CTAB concentration of 7×10^{-4} M, slightly below the CMC of CTAB in pure water of 9.5×10^{-4} M. In the presence of NaCl as the electrolyte, Pashley and Israelachvili (*38*) found a maximum contact angle on mica of about 65°.

However, they believe the accurate value to be closer to 95°, attributing the lower value to impurities removable by recrystallization from a 70:30 acetone-ether mixture (*39*). Yet, our work showed no significant difference in values for the surface tension or the contact angle when the purification of CTAB was carried out by recrystallization from a mixture of methanol and acetone followed by recrystallization from hexane saturated with ethanol. A possible explanation of the discrepancy between the values is the difference in the experimental procedures. Pashley et al.(*39*) measured the contact angle made by a drop of CTAB solution in air, whereas we measured the contact angle made by a captive air bubble inside the CTAB solution. Thus, we measured the water-receding angle first followed by the water-advancing angle, while the order was reversed in the earlier study by Pashley. Furthermore, in our study, the quartz slide was allowed to soak in CTAB solution prior to making the measurements.

In the case of the receding angle, our measured maximum value of near 50° is close to the value obtained by both McCaffery and Mungan (*37*) and Pashley et al. (*39*).

Capillary Pressure-Saturation Measurements. Figure 6 shows the drainage and imbibition capillary pressure-saturation measurements for various concentrations of CTAB at pH = 6. The imbibition relationship shows a greater range of behavior than the drainage relationship, implying a greater dependence on the surface conditions. At small contact angles, the aqueous phase imbibes strongly. As the contact angle increases, the aqueous phase imbibes less. At the isoelectric point of the CTAB-coated silica, corresponding to the maximum value of the contact angle of 64°,

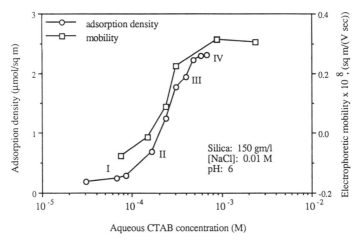

Figure 3. Adsorption density and electrophoretic mobility for CTAB on silica

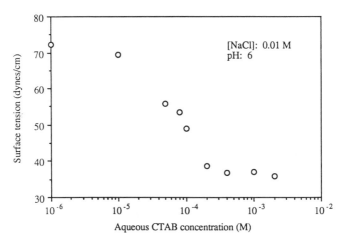

Figure 4. Surface tension of water as function of CTAB concentration

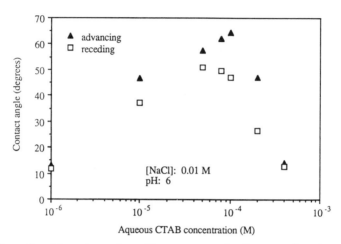

Figure 5. Advancing and receding contact angle of water as function of CTAB concentration

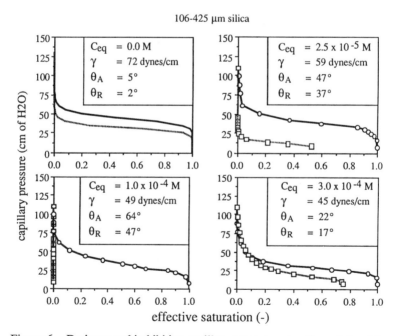

Figure 6. Drainage and imbibition capillary pressure-saturation relationships for various CTAB concentrations (pH=6, 0.01 M NaCl)

the aqueous phase does not imbibe. As the silica surface becomes positively charged, the contact angle decreases and the aqueous phase once again imbibes.

To determine whether these changes can be predicted on the basis of equation 9, the appropriate scaling factors were calculated (Table II). The intrinsic angles for drainage were the receding angles given in Figure 5, whereas the intrinsic angles for imbibition were the advancing angles shown in the same figure. These angles were corrected for roughness using the equations listed in Table I, and the suitable curvature correction factors, $Z(\theta)$, calculated according to the method outlined in Melrose (*20*). By multiplying the capillary pressures of the air-water-silica system by the calculated scaling factors, the corresponding capillary pressures for the air-water-silica-CTAB systems were estimated.

A comparison of the measured and estimated capillary pressure relationships is shown in Figure 7. This comparison demonstrates the methodology outlined here (and described in more detail in Demond and Roberts (*12*)) for organic liquid-water systems has the capability of predicting the effect of surfactant concentration on both drainage and imbibition capillary pressure-saturation relationships.

Table II. Calculated Scaling Factors

Process	CTAB conc. (moles/liter)	γ (dynes/cm)	θ_{INT} (degs)	θ_r (degs)	$Z(\theta)$	Scaling factor
DRAINAGE	0.0	72	2	0	1.00	1.00
	2.5×10^{-5}	59	37	2	1.05	0.86
	1.0×10^{-4}	49	47	4	1.12	0.68
	3.0×10^{-4}	45	17	0	1.01	0.60
IMBIBITION	0.0	72	5	0	0.94	1.00
	2.5×10^{-5}	59	47	51	0.21	0.12
	1.0×10^{-4}	49	64	85	No imbibition	0.00
	3.0×10^{-4}	45	22	1	0.68	0.31

Conclusion

The results of this research show the relationship between the sorption of CTAB and the capillary pressure-saturation relationships for an air-water-silica system. As CTAB sorbs to the silica surface, the silica surface becomes less negatively charged. This is reflected in the increase in the contact angle, and the corresponding decrease in spontaneous imbibition in the capillary pressure measurements. At the isoelectric point of the silica surface, the contact angle passes through a maximum of about 65°, corresponding to a total lack of imbibition. As a bilayer of surfactant molecules is formed on the silica surface, the surface becomes positively charged; the contact angle

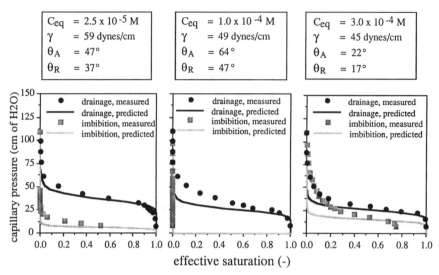

Figure 7.   Comparison of measured and predicted capillary pressure-
saturation relationships (pH=6, 0.01 M NaCl)

decreases, and substantial imbibition occurs again. A comparison of the measured and
calculated capillary pressure relationships for these systems shows reasonable
agreement, thus suggesting that the effect of surfactant sorption on capillary pressure-
saturation relationships can be estimated if the surface tensions and contact angles are
known. Because of the magnitude of the impact of surfactant sorption on the capillary
pressure-saturation relationships, it is anticipated that the presence of surface-active
contaminants in the vadose zone will have a significant effect on the advective
movement of water and its dissolved constituents.

Symbols and Abbreviations

a_i activity of species, i, in solution
C   concentration
CMC   critical micelle concentration
CTAB   cetyltrimetylammonium bromide
k_r relative permeability
K   unsaturated hydraulic conductivity
K_{sat} saturated hydraulic conductivity
n   porosity
P   pressure head
r   radius of pore
R   principal radius of curvature
R   gas constant
S   saturation
S_e effective saturation
S_r residual saturation
T   temperature
$Z(\theta)$ curvature correction factor

Greek Symbols:
γ   interfacial tension
Γ_i sorption density of species, i, at a given interface
θ   contact angle
θ   moisture content of wetting phase

Subscripts:
A   advancing
AIR   air phase
c   capillary
eq   equilibrium
H_2O aqueous phase
INT   intrinsic
NW   nonwetting
r   corrected for roughness
R   receding
S   solid
W   wetting

Acknowledgments

We gratefully acknowledge the support of Dr. Frank J. Wobber, Program Manager, Subsurface Science Program, Office of Health and Environmental Research, U.S. Department of Energy (Grant DOE-FG02-89-ER60820) for funding this research. In addition, we thank Haza Hammad for his contributions to the surface chemical measurements, Linda Timmer Verhulst for the surface tension and contact angle measurements, and Benjamin Witherell for the capillary pressure-saturation measurements.

Literature Cited

1) Arya, L.M.; Paris, J.F. *Soil Science Society of America Journal* **1981**, *45*, 1023-1030.
2) Haverkamp, R.; Parlange, J.Y. *Soil Science* **1986**, *142(6)*, 325-339.
3) Luckner, L.; van Genuchten, M.; Nielsen, D.R. *Water Resources Research* **1989**, *25(10)*, 2187-2194.
4) Ginn, M.E. In *Cationic Surfactants*, Jungermann, E., Ed., Surfactant Science Series, Marcel Dekker, NY, 1970, Vol. 4, 341-367.
5) Aronson, M.P.; Princen, H.M. *Colloid and Polymer Science* **1978**, *256*, 140-149.
6) McGuiggan, P.M.; Pashley, R.M. *Colloids and Surfaces* **1987**, *27*, 277-287.
7) Fuerstenau, D.W.; Herrera-Urbina, R. In *Cationic Surfactants*, Rubingh, D.N. and Holland, P.M., Eds., Surfactant Science Series, Marcel Dekker, NY, Vol. 37, 407-447.
8) Buckley, J.S.; Takamura, K.; Morrow, N.R. *SPE Reservoir Engineering* **1989**, *4(3)*, 332-340.
9) Gaudin, A.M.; Decker, T.G. *Journal of Colloid and Interface Science* **1967**, *24*, 151-158.
10) Morrow, N. R. *Journal of Canadian Petroleum Technology* **1976**, *15(4)*, 49-69.
11) Morrow, R.N.; McCaffery, F.G. In *Wetting, Spreading, and Adhesion*, Padday, J.F., Ed., Academic Press, New York, 1981, 387-411.
12) Demond, A.H.; Roberts, P.V. *Water Resources Research* **1991**, *27(3)*, 423-437.

13) U.S. Department of Energy. In *Subsurface Science Program: Program Overview and Research Abstracts FY 1989 - FY 1990*, U.S. Department of Energy Report No. DOE/ER-0432, Office of Energy Research, Office of Health and Environmental Research,Washington D.C., 1990, 5.
14) Milly, P.C.D. *Transport in Porous Media* **1988**, *3*, 491-514.
15) Celia, M.A.; Bouloutas, E.T.; Zarba, R.L. *Water Resources Research* **1990**, *26*, 1483-1496.
16) Lenhard, R.J.; Parker, J.C. *Journal of Contaminant Hydrology* **1987**, *1*, 407-424.
17) Lenhard, R.J.; Parker, J.C. *Water Resources Research* **1987**, *23(12)*, 2197-2206.
18) Demond, A.H.; Roberts, P.V. *Water Resources Research* **1991**, submitted for publication.
19) Shephard, J.W.; Bartell. F.E. *Journal of Physical Chemistry* **1953**, *57*, 458-463.
20) Melrose, J.C. *Society of Petroleum Engineers Journal* **1965**, *5*, 259-271.
21) Morrow, N.R. *Journal of Canadian Petroleum Technology* **1975**, *14(4)*, 42-53.
22) Zografi, G.; Johnson, B.A. *International Journal of Pharmaceutics* **1984**, *22*, 159-176.
23) Leverett, M.C. *Transactions of AIME* **1941**, *142*, 152-169.
24) Kunze, G.W.; Dixon, J.B. In *Methods of Soil Analysis, Part 1: Physical and Mineralogical Methods*, Klute, A., Ed.; American Society of Agronomy: Madison, WI, 1986, 95.
25) Mukerjee, P.; Mysels, K.J. *Critical Micelle Concentrations of Aqueous Surfactant Solutions*; National Standards Reference Data; National Bureau of Standards: Washington, DC, 1971.
26) MacRitchie, F. *Chemistry at Interfaces*; Academic Press: San Diego, CA, 1990.
27) O'Brien, R.W. *Journal of Fluid Mechanics* **1988**, *190*, 71-86.
28) Cheng, P.; Li, D.; Boruvka, L.; Rotenberg, Y.; Neumann, A.W. *Colloids and Surfaces* **1990**, *43*, 151-167.
29) Rotenberg, L.; Boruvka, L.; Neumann, A.W. *Journal of Colloid and Interface Science* **1983**, *93(1)*, 169-183.
30) Reginato, R.J.; van Bavel, C.H.M. *Soil Science Society of America Proceedings* **1962**, *26(1)*, 1-3.
31) Rosen, M.J.; Goldsmith, H.A. *Systematic Analysis of Surface-active Agents*; Wiley Interscience: New York, NY, 1972; 427-29.
32) Parks, G.A. *Chemical Reviews* **1965**, *65*, 177-198.
33) Bijsterbosch, B.H. *Journal of Colloid and Interface Science* **1974**, *47(1)*, 186-98.
34) Stratton-Crawley, R.; Shergold, H.L. *Colloids and Surfaces* **1981**, *2*, 145-154.
35) Rosen, M.J. *Surfactants and Interfacial Phenomena*; John Wiley: New York, **1989**.
36) McCaffery, F.G.; Cram, P.J. *Wetting and Adsorption Studies of the n-Dodecane-Aqueous Solution-Quartz System*; Research Report RR-13; Petroleum Recovery Research Institute: Calgary, Canada, 1971.
37) McCaffery, F.G.; Mungan, N. *Journal of Canadian Petroleum Technology* **1970**, *9(3)*, 185-196.
38) Pashley, R.M.; Israelachvili, J.N. *Colloids and Surfaces* **1981**, *2*, 169-187.
39) Pashley, R.M.; McGuiggan, P.M.; Horn, R.G.; Ninham, B.W. *Journal of Colloid and Interface Science* **1988**, *126(2)*, 569-578.

RECEIVED December 18, 1991

Chapter 12

Surfactant-Enhanced Solubilization of Tetrachloroethylene and Degradation Products in Pump and Treat Remediation

Candida Cook West

Robert S. Kerr Environmental Research Laboratory, U.S. Environmental Protection Agency, Ada, OK 74820

Experiments were conducted to investigate the enhanced solubilization of tetrachloroethylene (PCE), trichloroethylene (TCE), and 1,2-dichloroethylene (DCE) in nonionic surfactant solutions of Triton X-100, Brij-30, Igepal CA-720, and Tergitol NP-10 (alkyl polyoxyethylenes). Surfactant solubilization is being considered as a means to enhance mobile phase solubilities of ground-water contaminants for the purpose of improving the efficiency of pump and treat remediation. The primary objectives of this study were to observe the solubilization of relatively hydrophilic organic solutes at system temperatures similar to ground-water conditions and to determine if solubilization can be linearly correlated to the octanol/water partition coefficient, as has been observed by others for hydrophobic organic solutes. The results of this study show that surfactant solubilization of hydrophilic solutes is highly correlated with their octanol/water partition coefficient when corrected for temperature effects. It was also observed that there appears to be little difference in solubilizing efficiency between the four surfactants.

Remediation of contaminated ground-water at Superfund sites often entails extraction of contaminated ground-water with subsequent treatment of the extracted water until preset remediation clean-up levels are achieved. This remediation technology, commonly referred to as pump and treat, has been recognized as frequently being inefficient (1). Several factors are recognized as contributing to the failure of the technology to achieve health based clean-up criteria at studied sites. The non-aqueous phase liquids (NAPLs) present particular difficulties to remediate. NAPLs have entered many sites in sufficient quantities to deposit organic "ganglia," referred to as residual saturation, as they move through the vadose and saturated zones. Excess NAPL may also accumulate as bulk free phase organic liquid at interfaces. Both the residual saturation and bulk free phase represent long-term sources of contamination of water reservoirs. Dense non-aqueous phase liquids (DNAPLs), such as those studied here, represent a special case because of their unique physical and chemical properties. They are water soluble (ppm range) relative to their maximum contaminant levels (ppb range) and have high densities (Table I).

Table I. Formulas and properties of study DNAPLs

Compound	mol formula	MW	Density (g/ml)	Solubility[1] at 17°C ppm (mol/L)	Log K_{ow}[2] at 17°C
PCE	C_2Cl_4	167	1.63	239 (1.4 x 10^{-3})	3.14
TCE	C_2HCl_3	132	1.46	1385(1.0 x 10^{-2})	2.43
DCE	$C_2H_2Cl_2$	97	1.26	4761 (5.0 x 10^{-2})	1.83

[1] Determined in this study
[2] Estimated from log K_{ow}=-0.862 Log $S_{w,(mol/l),17°C}$+0.710 (6).

Currently there is interest in developing pump and treat enhancement techniques so that site remediation might be achieved in more reasonable time frames. This is particularly true for DNAPLs, since the mass of DNAPL deposited as residual saturation or bulk free phase can contaminate a correspondingly large volume of water at or below saturated concentrations resulting in long-term remediation efforts. One of the enhancement technologies that is being pursued involves the use of surfactants to increase the solubility of organic contaminants in the mobile phase (water/surfactant solution). The purpose of increasing the mobile phase solubilities of these contaminants is to increase the mass removed per volume of water pumped out of the subsurface for treatment. One mechanism of enhanced solubilization of these compounds occurs due to formation of micelles. The micelles act as an organic pseudophase which is dispersed in, and transports with, the mobile aqueous fluid. Micelles are organized surfactant structures that form spontaneously in solution when the surfactant concentration is above a level referred to as the critical micelle concentration (CMC). Below the CMC, surfactant molecules exist solely as monomers. Surfactant molecules added above the CMC orient themselves in aggregates so that, in aqueous solution, the polar or ionic portions of the molecules are pointed towards the bulk solution and the non-polar tails are oriented away from the bulk solution. The organic interior (core) of the micelle represents a volume of hydrocarbon that has a high capacity for solubilizing an organic contaminant, thus increasing the overall carrying capacity (solubilization potential) of the fluid phase. By definition, all surfactants are surface active though not all surfactants form micelles. The formation of micelles is evidenced by a sudden change in any one of a number of solution parameters at the CMC (surface tension, conductivity, etc.) indicating that a radical change in the aqueous surfactant environment has occurred. For those surfactants that do not form micelles, there is simply a linear change in solution parameters with increasing concentration of the surfactant with no apparent anomalies.

Recently, the effects of surfactants on the water solubility enhancement of hydrophobic organic contaminants have been reported. The mechanism of solubilization of these compounds can be conceptualized as a two-step process; the first being the solubilization by soluble monomers below the CMC and the second being solubilization by dispersed micelles above the CMC. Kile and Chiou (2) described the solubilization mechanisms above and below the CMC by the general expression:

$$S_w^*/S_w = 1 + X_{mn}K_{mn} + X_{mc}K_{mc} \qquad (1)$$

where S_{w^*} = solubility of the solute in the surfactant solution
S_w = solubility of the solute in pure water
X_{mn} = concentration of the surfactant as monomers
X_{mc} = concentration of the surfactant as micelles
K_{mn} = partition constant for the solute between water and monomers
K_{mc} = partition constant for the solute between water and micelles

S_w^*/S_w is the enhanced water solubility of the organic compound in surfactant solution relative to its solubility in pure water. $X_{mn}K_{mn}$ is the portion of the observed enhanced water solubility contributed by the cosolvency effect of the surfactant monomers below the CMC, and $X_{mc}K_{mc}$ is the portion contributed by incorporation of the compound into the micelles. In the study of Kile and Chiou (2), solubilization enhancement of DDT and trichlorobenzene (TCB) in Triton X-100, Triton X-114, Triton X-405, Brij 35, sodium dodecyl sulfate and cetyltrimethylammonium bromide was observed. They concluded that the water solubility enhancement by surfactant solutions below the CMC can be significant for compounds of extremely low water solubility such as DDT.

Kile and Chiou (3) also observed enhanced water solubility for DDT and TCB by petroleum sulfonates, which are commercial mixtures of emulsified sulfonated hydrocarbons and free mineral oils. These surfactant mixtures do not form micelles as evidenced by the linear decrease in surface tension for concentrations up to 500 mg/l. The water solubility enhancement of DDT and TCB in these surfactants solutions increased linearly with sulfonate concentration. In a study conducted by Edwards et al. (4) it was observed that the "apparent" solubility of polynuclear aromatic hydrocarbons was enhanced by a factor of two in nonionic surfactant solutions below the CMC.

The vast majority of surfactant solubilization studies have been conducted above the CMC and results reported using typical surfactant theory to describe observations. In some cases, this theory has been adapted to relationships derived in the environmental literature. For instance, the relationship between K_m (the ratio of the mole fraction of the solute in the micellar pseudophase (X_m) to the mole fraction of the solute in the monomer pseudophase (X_a)) and K_{ow}, the octanol/water partition coefficient, has been examined (4,5). K_m can be determined from the relationship:

$$X_m = MSR/(1 + MSR) \qquad (2)$$

where MSR = molar solubilization ratio, defined as the number of moles of the solute that can be solubilized per mole of the surfactant added to the solution. A complete derivation of this theory has been presented by Edwards et al. (4). The MSR for a surfactant/organic solute system can be determined from the slope of the plot of the concentration of the solute in solution versus the concentration of the surfactant above the CMC. The value of K_m may be calculated from equation (2) using the MSR and the observed solubility in the monomer solution. In this study, log K_m was determined for the three study compounds in four nonionic surfactants at 17°C and compared with their calculated log K_{ow}'s at 17°C to determine if a linear relationship exists between the two parameters.

Materials and Experimental Methods

Study Compounds and Surfactants. Trichloroethylene and tetrachloroethylene (both >98% purity) were obtained as ^{14}C-labeled compounds (specific activities 4.6 and 12.8 mCi/mmol, respectively). Radiolabeled TCE and PCE were separately washed from the containers with high purity non-labeled TCE (99+% spectrophotometric grade) and PCE (99.9% HPLC grade), respectively. The compounds were used neat in these experiments. Activities were related to concentrations by analyzing the samples using volatile organic analysis as described below. Non-labeled 1,2-dichloroethylene (98% trans isomer) was obtained and considered to be pure trans isomer for the purpose of these experiments. All solutes were purchased from Aldrich Chemical Co. Selected characteristics of these chemicals are listed in Table I. The surfactants were donated by the GAF Corp. (Igepal CA-720), Union Carbide Chemicals and Plastics Co. (Tergitol NP-10), ICI Americas, Inc. (Brij 30), and Rohm and Haas Co. (Triton X-100). There was no purity information supplied for the chemicals and they were used as received. Selected chemical characteristics for the surfactants are listed in Table II.

Table II. Surfactants employed in this study

Surfactant	av mol formula	av MW
Brij 30	$C_{12}H_{25}O(CH_2CH_2O)_4H$	363
Igepal CA-720	$C_8H_{17}C_6H_4O(CH_2CH_2O)_{12}H$	753
Tergitol NP-10	$C_9H_{19}C_6H_4O(CH_2CH_2O)_{10.5}H$	683
Triton X-100	$C_8H_{17}C_6H_4O(CH_2CH_2O)_{9.5}H$	625

Surface Tension Measurements. Surface tension measurements were made with a Fisher Model 20 du Nouy ring tensiometer. The instrument was calibrated daily with water and benzene. Corrections were made for the dial reading and ring geometry. All glassware was cleaned scrupulously and fired at 450°C prior to use. Measurements were made at 17°C by placing the beaker containing the surfactant solution into a jacketed beaker in-line with a constant temperature waterbath. The entire apparatus was placed in a glovebox to keep the surface of the solutions free of airborne dust or particulate matter. Measurements were made in triplicate and the average value reported. The CMCs determined by these measurements showed that the surfactants were consistent with those used in studies conducted by Edwards et al. (4).

Batch Studies. Batch solubilization tests were conducted for each of 12 solute/surfactant systems over a wide range of surfactant concentrations below and above the CMC of each surfactant. All aspects of the experiments were conducted at 17±1°C. Solubilization systems were comprised of 8.5 ml vials equipped with open top screw caps and teflon laminated discs (Tuf-bond from Pierce Co.). All glassware and septa were autoclaved. The distilled water used to prepare solutions was boiled and filter-sterilized to reduce the probability of microbiological activity. Surfactant solutions were prepared in batches from freshly prepared concentrated stock solutions to ensure precise concentrations across each set of surfactant concentrations. The vials were filled with the desired surfactant solution, leaving minimum headspace, and then the test DNAPL was added in sufficient volume to provide an excess above the compound solubility. Ten surfactant concentrations were prepared, ranging above and below the CMC (10^{-8} to 1% by weight). The vials were prepared in triplicate, as were appropriate blanks. All samples were equilibrated for two to five days in a temperature-controlled, dark chamber at 17±1°C. The equilibration time had been predetermined to be adequate. Excess DNAPL solute was separated prior to analysis by centrifuging the samples at 17°C at 1000 rpm for 10 minutes.

Analytical Methods. ^{14}C activities of each sample were determined on an LKB Model 1219 Rackbeta liquid scintillation counter employing automatic calibration. Activity/concentration curves were derived from gas chromatographic analysis of prepared radiolabelled standard solutions which were then used to convert activities to concentrations for each sample. Non-radiolabeled 1,2-dichloroethylene concentrations were determined by preparing subsamples for analysis using gas chromatography equipped with an FID detector. Fluorobenzene was employed as an internal standard for all samples. Sample concentrations were calculated as the mean of three replicates.

Results and Discussion

Solubilization Relationships. In these studies, it was observed that there was no detectable increase in the solubility of the solutes in the monomer solutions of the four nonionic surfactants as compared to pure water controls run for each surfactant/solute system. The mean solubilities and associated standard deviations of PCE, TCE, and DCE in pure water (based on twelve measurements for each solute) were 239±40, 1385±67, and 4761±246 ppm for PCE, TCE, and DCE, respectively. These results agree with the observations of Kile and Chiou (2) who observed nominal solubility enhancement of 1,2,3-trichlorobenzene in several nonionic and anionic monomeric surfactant solutions. The water solubility of the least soluble solute studied here (PCE) is approximately an order of magnitude greater than TCB. Solubilization of PCE, TCE, and DCE was plotted as a function of surfactant solution concentration for each of the four surfactants, and the MSRs were determined from the slope of the curves above the CMC. This is shown in

Figure 1 for DCE solubilization by Igepal CA-720 near the CMC. Table III reports the calculated MSR and log K_m for each surfactant/solute system. Log K_m for each solute in four surfactants are essentially the same (i.e. igepal/PCE≈tergitol/PCE ≈brij/PCE≈triton/PCE) indicating that within a general class of surfactants, there may be little variation in solubilizing efficiency. The solubilization appears to be more highly correlated with the water solubility of the solute (i.e. igepal/PCE >igepal/TCE>igepal/DCE).

S_{w*}/S_w at surfactant saturation was estimated when the solute concentration stabilized (generally around 1% by wt. surfactant)(See Figure 2). As shown by the results in Table IV, there was little variation in the S_{w*}/S_w value for each DNAPL/surfactant system. Within experimental error, the enhanced solubilization was approximately the same for each of the three chlorinated hydrocarbons.

Log K_m-log K_{ow}. Figure 3 and Table V show the correlation between log K_m and log K_{ow} for the four surfactant solutions. K_m and K_{ow} values were calculated using the water solubilities determined at 17°C. The slopes for each surfactant set were not significantly different from one another, since essentially all the data points were within one standard deviation of the mean of the pooled slopes. These results compare quite well with those of Edwards et al. (4) and Valsaraj and Thibodeaux (5) shown in Figure 4. Edwards et al. (4) used the same surfactants utilized in this study with sparingly soluble polynuclear aromatic compounds; Valasaraj and Thibodeaux (5) used sodium dodecylsulfate with both soluble and sparingly soluble compounds. In each case, a linear relationship was observed. The most significant differences between the results of these studies were the intercepts, which suggests some difference in solubilization efficiency for different classes of surfactants (nonionic versus anionic).

Conclusions. The apparent solubilities of PCE, TCE, and DCE were measured in solutions of four nonionic alkyl polyoxyethylene surfactants. The apparent solubilities of these solutes were unchanged in monomeric surfactant solutions as compared to pure water solubilities. The molar solubilization ratio (MSR) and micellar-phase/aqueous phase partition coefficients (K_m) were calculated for each solute/surfactant system. Log K_m values were in good agreement with those predicted by others who have studied more hydrophobic organic solutes in the same surfactant solutions (4). Measurements of maximum solubilization for the PCE, TCE, and DCE in saturated surfactant solutions indicate that the pure water solubility of the solute may be a good estimator of solubilization. Log K_m values correlated very well to octanol/water partition coefficients (log K_{ow}), as long as the aqueous solubility and K_{ow} were corrected for temperature effects. Studies such as these will be useful in determining which parameters will be important in developing effective surfactant systems as candidates for the enhancement of pump and treat remediation.

Disclaimer

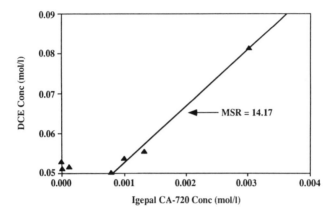

Figure 1. Determination of the MSR from the solubilization data of DCE in Igepal CA-720 near the CMC.

Table III. DNAPL Solubilization

Surfactant/Organic		MSR	log K_m*
Brij 30	PCE	1.34	4.33
	TCE	1.39	3.51
	DCE	5.17	2.97
Igepal CA-720	PCE	0.59	4.14
	TCE	3.43	3.63
	DCE	14.17	3.01
Triton X-100	PCE	3.97	4.47
	TCE	1.23	3.49
	DCE	6.66	2.98
Tergitol NP-100	PCE	3.21	4.45
	TCE	15.18	3.72
	DCE	**	**

* where $K_m = [55.4/S_{DNAPL,cmc}][MSR/(1+MSR)]$ (4) and the solubilities of the DNAPLs are 1.4e-3, 1.0e-2, and 5.0e-2 mol/l for PCE, TCE, and DCE, respectively, as determined in this study.
**unable to determine due to system instability.

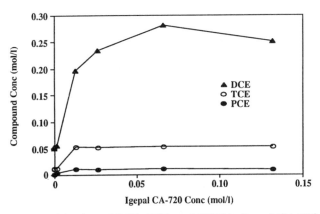

Figure 2. Solubilization of PCE, TCE, and DCE in Igepal CA-720 at 17°
Celsius.

Table IV. Chlorinated hydrocarbon solubilization in saturated surfactant
         solutions at 17°C

Surfactant/Organic		S_w^*/S_w
Brij 30	PCE	5.6
	TCE	4.6
	DCE	3.9
Igepal CA-720	PCE	9.7
	TCE	5.6
	DCE	5.6
Triton X-100	PCE	7.5
	TCE	4.4
	DCE	4.3
Tergitol NP-100	PCE	5.6
	TCE	4.6
	DCE	**

**unable to determine due to system instability.

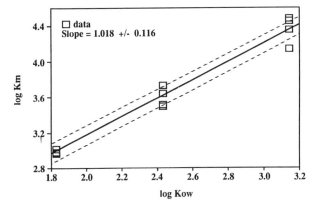

Figure 3. Comparison of log K_m-log K_{ow} correlations for PCE, TCE, and DCE in the four nonionic surfactant solutions. The slopes for each surfactant are listed in Table 11.5. The solid line refers to the average of the slopes and the dashed line indicates ± one standard deviation.

Table V. Relationship between log K_{ow} and log K_m for PCE, TCE, and DCE in four nonionic surfactant solutions

Surfactant	Slope	Intercept	R-squared
Igepal CA-720	1.04	+1.03	0.995
Brij 30	0.86	+1.48	0.989
Tergitol NP-10	1.03	+1.22	**
Triton X-100	1.14	+0.82	0.983

X_{slope}= 1.018±0.116; range= 0.902-1.134.
**insufficient data to determine

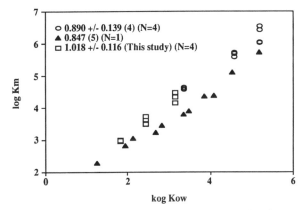

Figure 4. Comparison of log K_m-log K_{ow} correlations. The slope and standard deviation for the correlation observed in each study are listed in the legend. N refers to the number of surfactant systems studied.

Literature Cited

(1) Haley, J.L.;Hanson, B.;Enfield, C.;Glass, J. *Ground Water Monitoring Review.* *Winter* 1991, 119-124.
(2) Kile, D.E.;Chiou, C.T. *Envir. Sci. Technol.* 1989, *23*(7),832-838.
(3) Kile, D.E.;Chiou, C.T.;Helburn, R.S. *Envir. Sci. Technol.* 1990, *24*(2), 205-205.
(4) Edwards, D.A.;Luthy, R.G.;Liu, Z. *Envir. Sci. Technol.* 1991, *25*(*1*), 127-133.
(5) Valsaraj, K.T.;Thibodeaux, L.J. *Wat. Res.* 1989, *23*(2), 183-189.
(6) Chiou, C.T.;Schmedding, D.W.;Manes, M. *Envir. Sci. Technol.* 1982, 16(*1*), 4-10.

RECEIVED January 8, 1992

Chapter 13

Solubilization and Biodegradation of Hydrophobic Organic Compounds in Soil–Aqueous Systems with Nonionic Surfactants

David A. Edwards, Shonali Laha, Zhongbao Liu, and Richard G. Luthy

Department of Civil Engineering, Carnegie Mellon University, Pittsburgh, PA 15213

Nonionic surfactants in soil/aqueous systems may desorb and solubilize HOCs and affect the course of HOC biodegradation at bulk solution surfactant concentrations greater than a critical micelle or aggregate concentration. Various issues need to be addressed with respect to the transport and fate of nonionic surfactants in soil/aqueous systems including: the effects of soil components on sorption, the kinetics of surfactant sorption, and the biodegradation of sorbed surfactants. The inhibitory effects of surfactant solubilization on biodegradation of HOCs need to be explored from both physicochemical and microbial perspectives.

Nonionic surfactants are a class of surface-active compounds that may strongly interact with hydrophobic organic compounds (HOCs), soil and microorganisms in soil/aqueous systems. These interactions affect the potential for surfactant-facilitated HOC transport in soil and groundwater systems, and the feasibility of engineered surfactant cleanup of contaminated sites (1). In freshwater systems at 25 C at bulk liquid surfactant concentrations ranging up to several orders of magnitude greater than a surfactant-specific critical concentration, most nonionic surfactants form regular micelles in single-phase solutions, whereas certain surfactants, such as $C_{12}E_4$, may form bilayer lamellae or other types of aggregates in more complex two-phase solutions (2, 3). The critical concentrations for the onset of micelle and aggregate formation are termed the critical micelle concentration (CMC) and the critical aggregation concentration (CAC), respectively. Important changes may occur in surfactant sorption, surfactant solubilization of HOCs, and microbial mineralization of HOCs in the presence of nonionic surfactants at or near these critical surfactant concentrations.

This chapter summarizes recent research at Carnegie Mellon University involving both laboratory experiments and modeling relevant to physicochemical and biological processes affecting HOCs in systems of soil and dilute nonionic surfactant solutions (4, 5, 6; Edwards et al., 1992a, 1992b, in publication; Liu et al., 1991, in publication). The nonionic surfactants employed in past and present studies include an alkylethoxylate ($C_{14}E_4$), which, like other alkylethoxylates, is relatively nontoxic, and

0097–6156/92/0491–0159$06.00/0

three alkylphenol ethoxylates, $C_8PE_{9.5}$, C_8PE_{12}, and $C_9PE_{10.5}$, which are useful in laboratory investigations but because of their degradation products may not be viable for field remedial applications. Although the scope of this chapter does not permit review of other recently-published surfactant research in the fields of environmental chemistry, microbiology, and engineering (e.g., 7,8,9,10,11,12), it does allow identification of some of the current research needs relating to surfactants in soil/aqueous systems.

Experimental Methods and Materials

The nonionic surfactants employed in these studies were used as received from the manufacturer or chemical distributor. Naphthalene, anthracene, phenanthrene, and pyrene are polycyclic aromatic hydrocarbons (PAHs) used in the studies as model HOC compounds. ^{14}C-labeled PAH compounds were obtained from Amersham Corporation or from Sigma Chemical Co., and nonlabeled PAH compounds were obtained from Aldrich Chemical Co., with purities greater than 98%. Stock solutions of PAH compounds contained known ratios of radiolabeled to nonlabeled PAH mass. The two soils employed in the various soil/aqueous experiments were an undisturbed Morton subhumid grassland soil from North Dakota, and a pristine A-horizon Hagerstown silt loam from the Agricultural Experimental Station at Pennsylvania State University. The aqueous solutions in the soil/aqueous systems were formulated with BOD water (13) with 0.02 M $Ca(NO_3)_2$ added to assist in the separation of soil solids from solution.

In solubilization tests with soil, batch test soil/aqueous samples with nonionic surfactant and PAH in 50 ml centrifuge tubes were rotated on a tube rotator periodically to maintain the soil in suspension during equilibration. The systems were centrifuged prior to the sampling of liquid supernatant in order to eliminate experimental artifacts arising from the presence of colloids. Aliquots from solution samples were expressed through preconditioned 0.22 μm teflon filters to further reduce soil-derived colloidal substances. The aliquots were each mixed with 10 ml of liquid scintillation cocktail (Scintiverse II, Fisher Scientific). The radioactivity of each solution containing a ^{14}C-labeled PAH compound was measured with a Beckman LS 5000 TD liquid scintillation counter (LSC) that employed H# quench monitoring and automatic quench compensation technique (5,6).

The extent of PAH solubilization in nonionic surfactant solution without soil was assessed for each surfactant-PAH combination in aqueous batch tests as a function of surfactant dose. The PAH stock solutions were formulated such that 20 to 80 times the PAH mass needed for aqueous saturation in each sample would be present. 8 ml vials containing 5 ml samples with PAH and varied surfactant concentrations were capped with open-port screw caps lined with Teflon-lined septa and placed in a 25 C water bath to be reciprocated at 80 cpm for about 24 hours. Duplicate aliquots were withdrawn by glass syringe from each sample and each filtered through a preconditioned 0.22 μm Teflon membrane into 10 ml of scintillation cocktail.

Nonionic surfactant sorption onto soil was evaluated for each surfactant at varied sub-CMC aqueous-phase concentrations by employing a DuNuoy ring apparatus to measure the surface tension of bulk nonionic surfactant solutions from both aqueous and soil/aqueous systems. Supra-CMC sorption of nonionic surfactant was assessed with either azo dye solubilization and spectrophotometric analysis or by measurement of chemical oxygen demand (COD), from which the amount of surfactant in bulk solution could be inferred (Liu et al., 1991, in publication).

Aerobic microbial mineralization of phenanthrene was followed by observing the evolution of $^{14}CO_2$ from a soil/aqueous system comprising BOD water, 10 g of air-dried and sieved (2 mm) Hagerstown soil spiked with phenanthrene solution, and a consortium of PAH-degrading bacteria (RET-PA-101) that previously had been isolated from PAH-contaminated soil. The system containing the bacteria was housed in a side-arm biometer flask with the side arm containing NaOH to trap evolved CO_2. The flasks were placed on ganged magnetic stirrer and stirred periodically to keep the soil in suspension. The activity of the evolved $^{14}CO_2$ present in the NaOH was measured by LSC counting and converted to percent mineralization (5).

Experimental and Modeling Results

Sorption of Nonionic Surfactant and HOCs onto Soil. Sorption of nonionic surfactant onto soil diminishes the amount of surfactant available for solubilization and transport of HOC in porous soil media. Sorption of surfactant onto soil at a particular surfactant dose can be determined by comparing surface tension measurements in aqueous and in soil/aqueous systems. A significant fraction of nonionic surfactant sorbs onto soil at sub-CMC aqueous-phase concentrations. For instance, in a soil/aqueous system consisting of 6.25 g of Morton soil, 0.045 L of BOD water, and the minimum amount of Tergitol NP 10 ($C_9PE_{10.5}$) required for the aqueous-phase CMC to be attained, approximately 97% of the nonionic surfactant is sorbed onto soil. Thus, the mass of nonionic surfactant that must be added before the aqueous-phase CMC is attained in a soil/aqueous system is much greater than that in an aqueous system without soil. The extent of sub-CMC sorption can be characterized with Freundlich isotherms. For the several micelle-forming nonionic surfactants and soils tested, micelles do not appear to sorb onto soil, and the amount of surfactant that is sorbed thus plateaus at a maximum value at the aqueous-phase CMC (Liu et al., 1991, in publication). For the lamellae-forming surfactant tested, however, maximum sorption apparently occurs at a bulk solution surfactant concentration greater than the CAC.

Several of these characteristics are illustrated in Figures 1 and 2. Figure 1 shows a plot of surface tension as a function of the logarithm of $C_8PE_{9.5}$ nonionic surfactant dose in an aqueous system and in a soil/aqueous system with about a 1:8 ratio of soil weight to water volume. The initiation of micelle formation is indicated by the minimal surfactant dose at which surface tension ceases to decline. The inflection point represents the approximate CMC or CAC, depending on the type of surfactant, beyond which point the surfactant monomer concentration remains essentially constant (7). The amount of surfactant sorbed at sub-CMC doses can be computed from the difference in dose necessary to attain a specified value of surface tension. Measurement of the surfactant concentration itself, or a micellar property, is necessary to assess surfactant sorption at doses that result in exceedance of the aqueous-phase CMC or CAC. Figure 2 shows a sorption isotherm for $C_8PE_{9.5}$ micelle-forming nonionic surfactant for sub-CMC and supra-CMC bulk solution surfactant. The figure shows a maximum, plateau value of sorption attained when the solution surfactant concentration attains the CMC.

Nonionic Surfactant Solubilization of HOCs. In aqueous or soil/aqueous systems, hydrophobic organic compounds can be solubilized within the hydrophobic interiors of nonionic micelles, or in the adjacent palisade layer, increasing the apparent bulk solution HOC solubility relative to the HOC solubility in pure water. Micellar solubilization in the bulk solution is initiated for PAHs and many other HOCs at the aqueous-phase CMC or CAC. The bulk solution may be considered for the purpose of modeling solubilization as consisting of two separate pseudophases: i) a micellar pseudophase comprising the hydrophobic interior portions of the surfactant micelles collectively, and ii) an aqueous pseudophase external to the micelles that consists of aqueous solution saturated with surfactant monomers. The partitioning of HOC between the micellar pseudophase and the aqueous pseudophase can be quantified with a partition coefficient, K_m, which is the mole fraction of HOC in the micellar pseudophase divided by the mole fraction of HOC in the aqueous pseudophase.

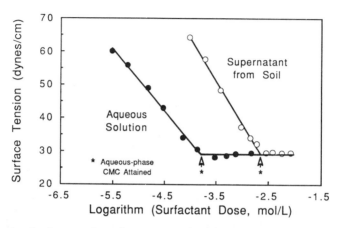

Figure 1.   Surface tension of aqueous and soil/aqueous systems with $C_8PE_{9.5}$ nonionic surfactant.

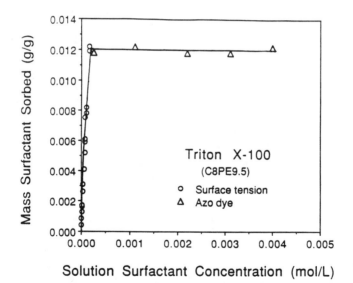

Figure 2.   Soil sorption isotherm for $C_8PE_{9.5}$ nonionic surfactant for sub-CMC and supra-CMC bulk solution surfactant.

Figure 3 shows a plot of phenanthrene solubilization by C_8PE_{12} nonionic surfactant in the absence of soil and in the presence of pure-phase phenanthrene. Phenanthrene solubilization commences at the CMC and is linear for surfactant concentrations greater than the CMC. The slope of this relationship is the molar solubilization ratio, MSR, with units of moles of HOC solubilized per mole of micellar surfactant. A value of K_m, which is applicable to systems in either the presence or absence of pure-phase HOC, can be calculated for each surfactant/HOC combination by using an experimental value for the MSR (4).

Figure 4 shows solubilization of pyrene by various surfactants at approximately 1:8 ratio of soil weight to water volume (6). The initial amount of pyrene in the soil/aqueous system without surfactant was selected so as to result in pyrene concentration at maximum aqueous solubility with no separate-phase HOC. These results show, among the surfactants evaluated, that the nonionic ethoxylate surfactants exhibited the best solubilization behavior. Solubilization for each nonionic surfactant appears to commence upon attainment of the CMC in the aqueous phase. Sodium dodecyl benzene sulfonate may have been ineffective in solubilizing pyrene owing to precipitation with 0.02 molar calcium in the test solution.

In a soil/aqueous system without separate-phase HOC, the bulk solution HOC apparent solubility in the presence of nonionic surfactant is a nonlinear function of surfactant dose. This overall nonlinear behavior, however, is simply a result of linear partitioning of HOC between the soil and the aqueous pseudophase, and between the aqueous pseudophase and the micellar pseudophase, as the number of micelles in the bulk solution increases (Edwards et al., 1992a, in publication). Supra-CMC nonionic surfactant bulk solution concentrations in such a system can result in substantial desorption of HOC from the soil by solubilization within surfactant micelles, which in porous media are colloids that are potentially mobile under hydrodynamic conditions. Surfactant monomers present in the aqueous pseudophase may affect the value of the HOC soil/aqueous-pseudophase partition coefficient, $K_{d,cmc}$, by enhancing the apparent HOC solubility in the aqueous pseudophase; the surfactant sorbed on soil, however, tends to increase the value of the partition coefficient by increasing the fractional organic carbon content of the soil. $K_{d,cmc}$ can be estimated from parameters for which values are known or can be estimated. The concentration of HOC in bulk solution and the fraction of HOC solubilized from soil can be modeled without calibration by employing model parameter values, e.g., K_m, $K_{d,cmc}$ and the surfactant sorption coefficient, obtained from independent experiments and/or estimation techniques. The results of this modeling approach are compared with experimental data in Figure 5 (Edwards et al., 1992b, in publication) for pyrene solubilized from soil as a function of dose of $C_8PE_{9.5}$ nonionic surfactant.

Nonionic Surfactant Effects on HOC Biodegradation. In soil/aqueous systems, highly hydrophobic organic compounds are typically found in the soil phase at concentrations much higher than in the aqueous phase, the extent of sorption being correlated with the soil organic fraction. The effect of HOC sorption on biodegradation is not well understood. The presence of soil sorbent may i) decrease HOC substrate utilization rate, due to desorption or diffusion limitation, or due to lowering aqueous-phase substrate concentration, or ii) increase the rate of mineralization due to immobilization of biomass and enhancing mass transfer from a substrate-enriched surface. In general it is believed that the effect of a sorbent solid on microbial activity is an indirect physicochemical feature influencing the substrate rather than the bacteria (14).

A number of HOC-degrading organisms produce emulsifying or solubilizing agents, and the dissolution and emulsification of such HOC compounds appears to have a positive effect on degradation rates. In certain laboratory tests and fermentation studies, the increase of aqueous solubilities of sparingly soluble substances by surfactant incorporation has been demonstrated to result in greater substrate

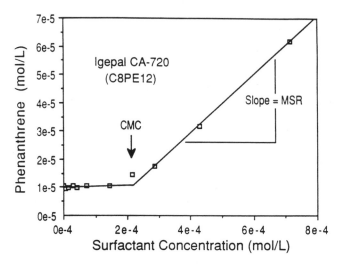

Figure 3. Solubilization of phenanthrene by C_8PE_{12} nonionic surfactant in aqueous system with excess phenanthrene.

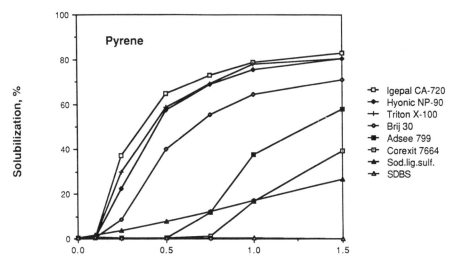

Figure 4. Solubilization of pyrene by various surfactants with a soil weight to water volume ratio of 1:8.

bioavailability for certain compounds (12,15,16). In a soil/aqueous system, the use of aqueous surfactant solutions may result in the transfer of HOC contaminant from the soil-sorbed phase to the micellar phase. For this reason, aqueous surfactant solutions have been suggested for use in soil bioremediation efforts to increase the bioavailability of hydrophobic organic compounds. With this in mind, the effects of nonionic surfactants on the biomineralization of the model hydrophobic organic compound, phenanthrene, in soil/aqueous systems was investigated. Results from such mineralization tests indicate that at sub-CMC surfactant levels the degradation of phenanthrene at best proceeds as fast as in the absence of surfactant.

Figure 6 shows the mineralization of phenanthrene in a soil/aqueous system as a function of time with varying concentrations of the nonionic surfactant $C_8PE_{9.5}$. At surfactant doses in excess of 0.05% (v/v) the microbial mineralization of phenanthrene is completely inhibited. This occurs at a surfactant dose of about 0.1%, a dose at which micelles form in the soil/aqueous system with the ratio of soil weight to water volume of about 1:8. The surfactant dose above which microbial mineralization of phenanthrene is inhibited thus appears to be related to the CMC of the surfactant in the presence of soil. Similar results were obtained for the nonionic surfactants $C_{12}E_4$ and $C_9PE_{10.5}$.

The inhibitory effect of higher surfactant doses on phenanthrene mineralization may be attributable to various phenomena, including toxic effects, preferential use of surfactant as substrate, lowering of aqueous-phase PAH concentration due to solubilization, or an interference of surfactant with microbial metabolic processes. Subsequent experiments evaluated some of these possible causes (5). Based on parallel tests with ^{14}C-glucose as the substrate, and more convincingly, on results from dilution experiments where supra-CMC surfactant doses were diluted to aqueous concentrations less than CMC, it appears that the microbial inhibition observed is not an irreversible toxicity effect.

Figure 7 shows the recovery of phenanthrene mineralization after several weeks following dilution of $C_8PE_{9.5}$ solution to a level below that resulting in aqueous-phase micelles. Experiments with glucose added as another substrate to the soil-water system indicate that although the presence of 0.13% (w/v) glucose produces a lag effect on the mineralization of phenanthrene, no dramatic inhibition occurs. Additional tests using higher concentrations of glucose are underway, and various other nonionic surfactants are being employed in sub- and supra-CMC doses in ongoing tests. Preliminary results appear to negate a surfactant-specific inhibitory effect.

In conclusion, sub-CMC doses of nonionic surfactant in soil/aqueous systems in the absence of separate-phase phenanthrene do not inhibit mineralization of phenanthrene, but neither do they enhance the degradation rate. At surfactant doses in excess of the aqueous-phase CMC, the nonionic surfactants exhibit an inhibitory effect on phenanthrene mineralization. The inhibition may be a result of a physicochemical effect of the surfactant micelles interfering with substrate transport into the cell, or with the activity of enzymes and other membrane proteins of the cell. The inhibition may also be a result of limited bioavailability of micellar phenanthrene due to slow exit rates from the micelles. The possibility of surfactant utilization as preferential substrate also needs to be examined in more detail.

Summary

Ongoing research is investigating mechanisms of nonionic surfactant sorption onto soil, solubilization of hydrophobic organic compounds (HOCs) from soil, and microbial degradation of HOCs in soil/aqueous systems with nonionic surfactants. Solubilization of HOCs by nonionic surfactant commences at a surfactant dose sufficient to attain the critical micelle concentration (CMC) or critical aggregate concentration (CAC) in the bulk solution. The sorption of surfactant onto soil results in the dose to achieve solubilization being much greater than that for an aqueous

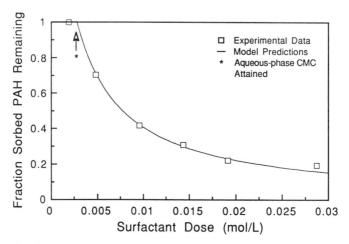

Figure 5.   Comparison of experimental data and model predictions for solubilization of pyrene from soil with $C_8PE_{9.5}$ nonionic surfactant.

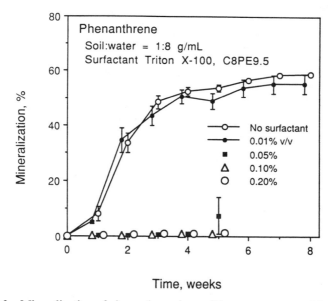

Time, weeks

Figure 6.   Mineralization of phenanthrene in a soil/aqueous system with varying concentrations of $C_8PE_{9.5}$ nonionic surfactant.

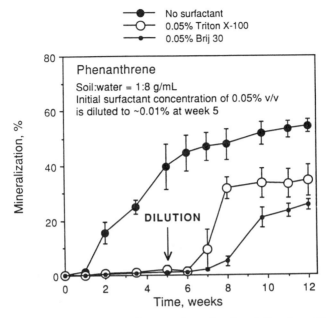

Figure 7. Recovery of phenanthrene mineralization following dilution of $C_8PE_{9.5}$ and $C_{12}E_4$ nonionic surfactants to sub-CMC bulk aqueous concentration.

system without soil. Regular micelle-forming nonionic surfactants appear to attain sorption maximum onto soil at a surfactant dose that results in the aqueous monomer concentration being equal to or slightly greater than the CMC; micelles in such systems appear not to sorb onto soil.

The solubilization of HOC from soil can be modeled using parameters obtained from independent experiments. The microbial degradation of phenanthrene in soil/aqueous systems is inhibited by addition of alkylethoxylate or alkyphenylethoxylate surfactants at doses which result in solubilization of phenanthrene from soil. Available data suggest that the supra-CMC inhibitory effect on biodegradation is not an irreversible toxic effect, or a consequence of the surfactant being used as a preferential substrate.

This work suggests various issues that need to be addressed with respect to transport and fate of nonionic surfactant in soil/aqueous systems, interaction between surfactant and HOCs, and possible applications to remediation technologies. With respect to sorption of nonionic surfactants onto soil it is not known how soil/aquifer components, i.e., humic matter, mineral matter, and clays, affect sorption. It is not known how the relative contribution of the surfactant hydrophobic and hydrophilic moieties affect sorption properties. The kinetics of surfactant desorption, and the biodegradation of sorbed surfactant, are important issues with respect to deployment of surfactants for engineered treatment of contaminated systems.

The effects of surfactant solubilization on biodegradation of HOCs need to be explored from both physicochemical and microbial perspectives. The effects of surfactant monomers and micelles on microbial cell surfaces and constituents must be better understood in order to evaluate whether synthetic surfactants may be employed advantageously to speed up rates of bioremediation. There is a need to know more of the physical and biochemical means by which surfactant may effect the transport of

HOCs to and across cell membranes. Clearly these various questions are difficult to address, requiring combined efforts of soil scientists, environmental engineers, and microbial ecologists.

Literature Cited

(1) Luthy, R.G.; Westall, J. In *Summary Report: Concepts in Manipulation of Groundwater Colloids for Environmental Restoration, Manteo, NC, October 15-18, 1990*; Editors: McCarthy, J.F. and Wober, F.J., U.S. Department of Energy: Oak Ridge Laboratory, TN., **1991**.

(2) Mitchell, D.J.; Tiddy, G.J.T.; Waring, L.; Bostock, T.; and McDonald, M.P. *J. Chem. Soc., Faraday Trans. 1*, **1983**, 79, pp. 975-1000.

(3) Rosen, M.J. *Surfactants and interfacial phenomena, 2nd ed.*, John Wiley and Sons, New York, **1989**.

(4) Edwards, D.A.; Luthy, R.G.; Liu, Z. *Environ. Sci. Technol.* **1991**, *25*, pp. 127-133.

(5) Laha, S.; Luthy, R.G. *Environ. Sci. Technol.* **1991**, *25*, pp. 1920-1930.

(6) Liu, Z.; Laha, S.; Luthy, R.G. *Water Sci. Tech.* **1991**, *23*, pp. 475-485.

(7) Chiou, G.T.; Kile, D.E.; Rutherford, D.W. *Environ. Sci. Technol.* **1991**, *25*, pp. 660-664.

(8) Abdul, A.S.; Gibson, T.L. *Environ. Sci. Technol.* **1991**, *25*, pp. 665-670.

(9) Jafvert, C.T; Heath, J.K. *Environ. Sci. Technol.* **1991**, *25*, pp. 1031-1038.

(10) Jafvert, C.T.*Environ. Sci. Technol.* **1991**, *25*, pp. 1039-1045.

(11) Holsen, T.M.; Taylor, E.R.; Seo Y.-C.; Anderson, P.R. *Environ. Sci. Technol.* **1991**, *25*, pp. 1585-1589.

(12) Aronstein, B.N.; Calvillo, Y.M.; Alexander, M. *Environ. Sci. Technol.* **1991**, *25*, pp. 1728-1731.

(13) APHA *Standard Methods for the Examination of Water and Wastewater*, 16th Ed., American Public Health Association, Washington, D.C., **1985**.

(14) van Loosdrecht, M.C.M.; Lyklema, J.; Norde, W.; Zehnder, A.J.B. *Microbiol. Reviews*, **1990**, *54*, pp. 75-87.

(15) Liu, D. *Wat. Res.* **1980**, *14*., pp. 1467-1475.

(16) Efroymson, R.A.; Alexander, M. *Appl. Environ. Microbiol.* **1991**, *57*, pp. 1441-1447.

RECEIVED December 18, 1991

Chapter 14

Sorption of Hydrophobic Organic Compounds and Nonionic Surfactants with Subsurface Materials

Candace Palmer[1], David A. Sabatini[1], and Jeffrey H. Harwell[2]

[1]School of Civil Engineering and Environmental Science and [2]School of Chemical Engineering and Materials Science, University of Oklahoma, Norman, OK 73019

The use of surfactants has been proposed for the characterization and remediation of contaminated ground water. However, little research has been conducted to analyze the movement of surfactants (especially nonionic surfactants--NISs) in the subsurface. The hypothesis of this research was that sorption of NISs is similar to sorption of NOCs and thus is a function of their hydrophobicity. Batch and column studies were conducted utilizing four neutral organic chemicals (NOCs--atrazine, diuron, o-xylene and p-xylene) and four alkylphenolpolyoxyethoxylate NISs (Igepal CA 620, CO 620, CO 630 and CO 660) with two subsurface materials (alluvial sand and sandstone). Sorption of NISs exceeded while NOCs agreed well with predictions based on Kow values. It appeared that the nature of the subsurface organic fraction ("mature" versus "recent") affected the sorption isotherms (linear versus nonlinear) for the NISs, and that they interacted with the mineral fractions of the media; neither of these were observed for the NOCs. Thus, the behavior of NISs differs from that of NOCs, requiring an improved fundamental understanding of sorption of NISs in the subsurface.

Organic chemicals are ubiquitous in our society and consequently are widespread ground water contaminants. Sorption is a major process affecting the migration of these organic chemicals in ground water. Much research has investigated the sorption and migration of neutral organic contaminants (NOCs) in ground water, including pesticides and petroleum hydrocarbons. As a result, the mobility of NOCs is fairly well understood. However, little research has investigated the sorption of nonionic surfactants (NISs, amphiphilic organics) in ground water. NISs may occur in ground water as a result of their widespread use in industrial products. Also, surfactants have been proposed for soil washing and enhanced

0097–6156/92/0491–0169$06.00/0

subsurface remediation. The focus of this research was to evaluate the sorption of a range of NISs with subsurface materials.

Literature Review

NOCs. Much research has been conducted in evaluation of the sorption of NOCs in the subsurface. The majority of this research initially centered on the sorption of pesticides with surface soils and subsurface media. Excellent reviews on this subject exist (*1, 2, 3*). More recent chemicals of interest include petroleum hydrocarbons, halogenated solvents, PCBs, etc., which commonly occur as ground water contaminants. In general, the sorption of NOCs in the subsurface environment is well understood.

Two fundamental motivations for sorption have been proposed based on the three elements involved in sorption; the solute (adsorbate, NOC), the solvent, and the solid surface (sorbent) (*4*). The first motivation is referred to as solvent motivated sorption. In this type of sorption, the accumulation of the NOC at the surface/interior of the media particles is motivated by the "dislike" of the NOC for the solvent (or, more accurately, the dislike of the solvent for the NOC). NOCs are nonpolar organics and are thus marginally soluble in water (hydrophobic). A second motivation for sorption is referred to as sorbent motivated sorption. In this case the solid phase (sorbent) has a greater affinity for the solute than does the solvent. Charged surfaces (e.g., clays) will attract ions to their surface in an attempt to neutralize the surface charge. This motivation may be active for ionic organics.

Sorption is commonly described by equilibrium isotherms (plots of q--mass of NOC sorbed per unit mass of media--versus equilibrium NOC concentration). In the case when this plot is linear (linear sorption isotherm), the slope of the line of this plot is indicative of the degree of sorption and is referred to as the linear equilibrium partition coefficient (K_p). If this plot is nonlinear, a nonlinear isotherm (Freundlich, Langmuir, BET) may be used. In general, for nonionic organics and for aquifer materials the linear isotherm has been found to be appropriate. Researchers have found that the level of sorption (as indicated by the magnitude of K_p) for a given pesticide for a number of materials is highly proportional to the organic content (f_{oc}) of the materials. Thus, while for a given organic the K_p value varied from soil to soil, if the K_p value was normalized by the f_{oc} the resulting parameter (K_{oc}) was much less variable between soils and was only a function of the NOC (and its hydrophobicity). Researchers have also attempted to correlate the K_{oc} value of a NOC to a fundamental property of the NOC and have found the octanol-water partition coefficient (K_{ow}) to be most successful. This parameter quantifies the partitioning of a chemical between an octanol (nonpolar) and water (polar) phase and thus is an indicator of the hydrophobicity of the chemical. Researchers have proposed empirical relationships to predict the K_{oc} values from values of K_{ow} of the NOC (e.g, *5, 6*). Thus, it is possible to estimate the K_p value of a NOC based on the K_{ow} of the NOC and the f_{oc} of the soil. However, it should be noted that this methodology assumes solvent motivated sorption and that the organic content is the major solid phase responsible for sorption. At very low levels of f_{oc} mineral surfaces can become the dominant player responsible for sorption.

NISs.   Surfactants (<u>surf</u>ace <u>act</u>ive <u>agents</u>) are comprised of hydrophobic (nonpolar) and hydrophilic (polar, ionic) moieties, which accounts for their surface activity. In an attempt to minimize the free energy of the system, surfactants prefer to accumulate at interfaces where the polar moiety can be in a polar phase and the nonpolar moiety can be in a nonpolar phase. Surfactants may be cationic, anionic or nonionic depending on the nature of the hydrophilic group. By varying the length of either the hydrophilic or hydrophobic moiety it is possible to control the extent to which the surfactant will accumulate in each phase. The surfactants may aggregate into thermodynamically favorable micelles above a critical concentration (known as the critical micelle concentration--CMC).

Numerous studies have addressed the sorption of NISs onto clay surfaces (*7, 8, 9*) and onto silica surfaces (*10, 11, 12*). Hampson et al. (*9*) used alkyl substituted polyoxyphenylsulfonamides (NISs) to determine the effect of varying ethoxy and alkyl groups on the sorption of the NISs with sand and clays. Increasing the ethoxy chain length increased the sorption while increasing the alkyl chain length appeared to decrease the apparent sorption. The use of NISs in chemical flooding processes has demonstrated that NISs partition into the oil phase (*13*).

Several recent studies have evaluated the sorption of NISs onto subsurface media (with organic and inorganic surfaces). Urano et al. (*14*) studied the sorption of an alkylpolyoxyethoxylate (6 alkyl groups and 6 ethoxylate groups) and an alkylphenolpolyoxyethoxylate (9 alkyl groups and 10 ethoxylate groups) with three soils ranging in organic content from 1.7% to 6.0%. A Freundlich isotherm was observed for the NISs (concentrations below the CMC) and the soils, with increasing sorption observed for increasing organic content. Abdul and Gibson (*15*) studied the sorption of an alkylpolyoxyethoxylate (Witconol SN70, 10 to 12 alkyl groups and 6 ethoxylate groups) with a sand having an organic carbon content of 0.5%. A Langmuirian isotherm was observed up to concentrations of 4000 mg/L, with a BET type isotherm observed above 4000 mg/L. Liu et al. (*16*) evaluated the sorption of four nonionic surfactants with a soil containing an organic content of 0.96%.   Three alkylphenolpolyoxyethoxylate NISs were evaluated as follows: Igepal CA-720 (8 alkyl and 12 ethoxylate groups), Tergitol NP-10 (9 alkyl and 10.5 ethoxylate groups), and Triton X-100 (8 alkyl and 9.5 ethoxylate groups). One alkylpolyoxyethoxylate was evaluted (Brij 30, 12 alkyl and 4 ethoxylate groups). The sorption of all four NISs was observed to follow a Freundlich isotherm at concentrations at or below the CMC.

Research Hypothesis and Objectives

The fundamental hypothesis of this research was that sorption of NISs with subsurface aquifer media is similar to sorption of NOCs and thus is a function of the hydrophobicity of the NIS. Three research objectives were identified to evaluate this hypothesis as follows: (1) to evaluate the batch sorption of NOCs and NISs (with varying levels of hydrophobicity) on two subsurface media; (2) to evaluate the chemical and media properties affecting the sorption of NISs in batch tests; and (3) to evaluate the mobility of NISs in continuous flow column studies.

Materials and Methods

Batch and column studies were conducted in order to evaluate the research objectives. Four NISs, four NOCs and two subsurface media were utilized in these studies. The materials and methods utilized are discussed below.

Aquifer Materials. Two aquifer materials were selected for evaluation in this study; Canadian River alluvium (CRA) and Garber Wellington sandstone (GWS). The alluvium is part of the Quaternary alluvium associated with flood deposition while the sandstone is part of the Permian Garber-Wellington formation, also known as the Central Oklahoma Aquifer. Table I shows fundamental properties of these two media.

Table I. Media Properties

Property	Alluvium (CRA)	Sandstone (GWS)
sand (%)	91	98
silt (%)	2	1
clay (%)	7	1
organic content (f_{oc})	0.0024	0.0016
cation exchange capacity (meq/100g)	9	2
hydraulic conductivity (ft/day)	4.7	6.3

Chemicals. Four NOCs (two pesticides and two petroleum hydrocarbons) and four NISs were evaluated in this research. The pesticides evaluated (atrazine and diuron) were obtained from Chem Service, Inc (Westchester, PA) and were 99% and 80% pure, respectively. The petroleum hydrocarbons investigated (o-xylene and m-xylene) were obtained from Fisher Scientific and were both reagent grade (99% pure). Fundamental properties of the NOCs are shown in Table II.

The NISs evaluated in this research were obtained from the GAF Corporation and were part of the Igepal group (alkylphenolpolyoxyethoxylates, $CH_3(CH_2)_mC_6H_4O(C_2H_4O)_nH$). The NISs have octyl and nonyl alkyl groups and range from seven to ten oxyethylene groups. Fundamental properties of the NISs are shown in Table III. It should be noted that NIS concentrations evaluated in this research were well below their CMC.

Table II. Fundamental Properties of NOCs

Property	Atrazine	Diuron	o-Xylene	m-Xylene
Molecular Formula	$C_8H_{14}ClN_5$	$C_9H_{10}N_5Cl_2$	C_8H_{10}	C_8H_{10}
Molecular Weight	216	233	106	106
Solubility (mg/L)	33	42	175	158
log K_{ow}	2.34	2.81	2.77	3.20
Vapor Pressure (mm Hg)	8×10^{-7}	9×10^{-6}	5	6

Table III. Fundamental Properties of NISs

Property	CA 620	CO 620	CO 630	CO 660
Molecular Weight	514	594	616	660
Alkyl Group	C8	C9	C9	C9
Ethylene Oxides	7	8.5	9	10
log K_{ow}[1]	3.54	2.96	2.76	2.37

[1] Estimated from Lyman et al. (*17*)

The NISs and NOCs were analyzed using reverse phase high performance liquid chromatography (RPHPLC). The stationary phase was a C_{20} alkyl chain attached to a silica base (partisil) and the mobile phase was 72 to 80% methanol to water (Baker Analyzed HPLC methanol and Nanopure water). UV detection was utilized with wavelengths of 231 nm for the NISs and atrazine, 271 nm for diuron and 212 for the xylenes.

Batch and Column Methods. Batch and column studies were conducted in this research to evaluate the research objectives. Batch adsorption experiments were conducted in 42 ml glass vials using a constant volume of solution (20 ml) at varying concentrations and a constant mass of aquifer material (10 g). A constant ionic strength was established by maintaining 0.01 N $CaCl_2$ in the reactor solutions. The batch studies were conducted for 24 hours at room temperature ($22^{o}C$) and included soil and solution blanks and duplicates. For the volatile hydrocarbons, the mass balance for determining sorption accounted for the hydrocarbons in the headspace. Oxidation of organic matter for select batch studies was accomplished by using 30% hydrogen peroxide, according to standard methods (*18*).

Laboratory column studies were performed using glass columns measuring 2.5 cm in diameter and with media lengths of 20 cm. The porosity of the media was 35% and the pore water velocity was maintained at 30 cm/hr using Masterflex pumps with silicone and teflon tubing. A conservative tracer (chloride) was utilized to determine the hydrodynamics of flow in the columns, resulting in a dispersion coefficient of 18 cm^2/hr. Column studies were conducted at room temperature (22°C).

Results and Discussion

The results of this research will be presented in three sections as follows: (1) sorption of NOCs in batch studies, (2) sorption of NISs in batch studies, and (3) columns studies.

Sorption of NOCs. Batch studies were utilized to determine the sorptive capacity of the two subsurface materials for the NOCs. The sorption isotherms for the NOCs were observed to be linear for both media (see Figure 1 for atrazine isotherm and Table IV for NOC isotherm data). Note that Freundlich coefficients are only shown in Table IV for isotherms that deviated from linearity. It is observed that sorption of the NOCs with the CRA was consistently higher than with the GWS (in keeping with the higher f_{oc} of the CRA). Normalization of the partition coefficient ($K_{oc} = K_p / f_{oc}$) is observed to reduce the parameter variability between media. It is also observed that K_{oc} values estimated from K_{ow} values agreed well with experimental K_{oc} values for the NOCs. These observations are in keeping with the current understanding of solvent motivated sorption of NOCs.

Sorption of NISs. The sorption isotherms for the NISs were observed to be linear for the GWS but nonlinear for the CRA (see Figure 2 for CO 620 isotherm and Table IV for NIS isotherm data). Note in Table IV that Freundlyich coefficients were only shown for the natural CRA medium (for the other cases the Freundlich exponent was one and K_{fr} would have the same value as K_p). Also note that a linear isotherm was fitted to all data (even the nonlinear data for the CRA medium); this was done to allow comparison of the resulting K_{oc} values with those estimated from K_{ow}. It is hypothesized that the nonlinearity of the NIS isotherms with CRA was due to the chemical immaturity of the organic matter in the CRA. Since the CRA is a recent deposit (as compared to the GWS), it is expected that the CRA material would not have weathered as much. This would suggest that the CRA medium has a broader distribution of components comprising the organic matter (still retains more of the polar functional groups) which may affect the nature of sorption of the NISs. These polar functional groups would not be expected to impact the sorption of neutral organic compounds significantly, thus explaining why the NOC isotherms were linear with the CRA. To test this hypothesis, the organic fraction of the CRA was oxidized and batch tests were conducted with CA 620 and CO 660. Figure 3 shows the

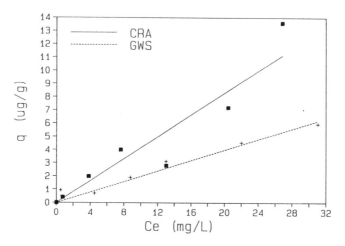

Figure 1: Atrazine Batch Isotherm

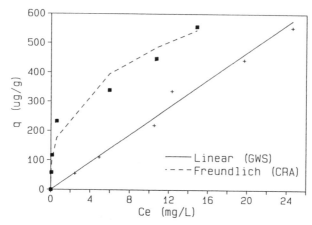

Figure 2: CO 620 Batch Isotherm

Table IV. Batch Data for NOCs and NISs

Chemical	Media	K_p (cm^3/g)	K_{oc} (cm^3/g)	$(K_{oc})est^*$ (cm^3/g)	$1/n$	K_{fr} $(\mu g/g)$
Atrazine	CRA	0.42	175	135	--	---
	GWS	0.20	125	135	--	---
Diuron	CRA	0.75	350	398	--	---
	GWS	0.56	312	398	--	---
o-Xylene	CRA	0.99	412	363	--	---
	GWS	0.65	406	363	--	---
m-Xylene	CRA	2.15	706	977	--	---
	GWS	1.13	896	977	--	---
CA 620	CRA	10.8	4,500	2138	0.54	59.3
	GWS	5.8	3,630	2138	--	---
	CRA-Ox	4.6	---	----	--	---
CO 620	CRA	41.1	17,400	562	0.38	194
	GWS	22.9	14,300	562	--	---
CO 630	CRA	31.2	13,000	355	0.48	117
	GWS	18.4	11,500	355	--	---
CO 660	CRA	19.7	8,210	145	0.44	99.7
	GWS	18.0	7,500	145	--	---
	CRA-Ox	17.2	---	---	--	---

Note: foc of media--CRA (0.0024), GWS (0.0016), CRA-oxidized (with hydrogen
       peroxide--below detection)
* Estimated from log K_{oc} = log K_{ow} - 0.21 (5)

resulting isotherm for CA 620 with the isotherm data shown in Table IV (CRA-Ox). It is observed that upon removal of the organic fraction the isotherms became linear, supporting the hypothesis stated above. Thus, it seems apparent that variations in the nature (possibly maturity) of organic matter can affect the sorption of amphiphilic compounds. A precedent for these observations can be found in two recent articles. Garbarini and Lion (19) concluded that inclusion of the sorbent's oxygen content in addition to the organic carbon content would improve the sorption description. Murphy et al. (20) found that adsorption of humics onto kaolinite and hematite resulted in differing sorptivities of the resulting surfaces for NOCs. This indicates that the same organic material associated with different mineral surfaces can result in differing sorptive characteristics.

It is also observed from Table IV that the levels of sorption of the NISs are significantly greater than that estimated based on hydrophobic theory (estimates based on K_{ow}). The NIS expected to be most strongly sorbed was observed to be the least strongly sorbed, but was the closest to the estimated level of sorption (CA 620). Within the nonyl series (CO 620, CO 630 and CO 660) the sorption of both media followed the anticipated trend of decreasing sorption with increasing oxyethylene groups, although the magnitude of sorption was 30 to 50 times greater than estimated. It is hypothesized that the sorption of NISs is affected by not only the organic fraction but also the mineral fraction of the media. As observed in Table IV, after the oxidation of the organic fraction, a significant level of sorption was still observed, supporting this hypothesis. Also, it would be expected that as the oxyethylene chain increases greater interactions with the mineral surfaces would be observed. This appears to be supported by the increasing ratio of measured to estimated K_{oc} values for increasing oxyethylene groups (e.g., for GWS and CO 620, CO 630, and CO 660, the ratio is 25, 32 and 52, respectively).

The difference in the sorption of CA 620 and CO 660 with the natural and oxidized CRA sample can be assumed to be the sorption of the NIS with the organic matter (total sorption minus the sorption with the mineral surfaces). For CA 620, the resulting K_{oc} value would be 2580 {(10.8-4.6)/0.0024}. This agrees well with the estimated K_{oc} value of 2138. However, for CO 660 the adjusted K_{oc} value would be 1040 {(19.7-17.2)/0.0024}, which is much greater than the estimated value of 145. This may be due to enhanced sorption by the mineral surfaces when the organic matter is present (an additive effect), due to errors in the estimation technique utilized in determining the hydrophobicity of CO 660, etc.

It is thus apparent that the sorption of amphiphilic organics is more complicated than the sorption of NOCs in aquifer materials. Certainly, the role of mineral surfaces is important with respect to the sorption of amphiphiles whereas they are generally not important in sorption of NOCs. Also, it is apparent that the nature of the organic matter affects the nature as well as the magnitude of the NIS sorption.

Columns Studies with NOCs and NISs. Two NOCs (atrazine and m-xylene) and two NISs (CA 620 and CO 660) were evaluated in column studies with both media. Figure 4 shows the breakthrough curves for CA 620 in both media. It is observed that the breakthrough curves are sigmoidal in shape, in keeping with chromatography theory. It is also observed that the CRA experienced more significant retardation relative to the GWS, consistent with batch results. Table V shows retardation factors for these NOCs and NISs as determined from the column runs and also as estimated from the batch studies. The decrease in retardation factors measured in column studies relative to that estimated from batch studies is assumed to be due to the kinetic limitations of continuous flow studies. Bouchard et al. (*21*) reported column to batch ratios of 0.43 to 0.74; the ratios in Table V fall into this range. The column results for CO 660 are somewhat anomalous in that the retardation in the columns was greater for the GWS than for the CRA. This is contrary to the batch results and the fact that the CRA has higher f_{oc}. It is hypothesized that this is due to either increased kinetic

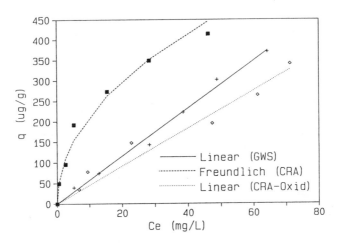

Figure 3: CA 620 Batch Isotherm (Including Oxidized CRA Sample)

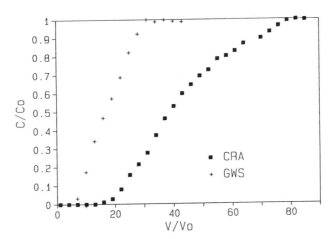

Figure 4: CA 620 Column Breakthrough Curves

Table V. Column Data for Selected NOCs and NISs

Chemical	Media	(Rf)column	(Rf)batch	(Rf)c/(Rf)b
Atrazine	CRA	1.7	2.8	0.61
	GWS	1.5	2.2	0.78
m-Xylene	CRA	7.8	12.0	0.65
	GWS	6.2	8.2	0.75
CA 620	CRA	38	50	0.76
	GWS	18	26	0.70
CO 660	CRA	38	85	0.44
	GWS	41	77	0.53

limitations of sorption with the less mature organic fraction associated with the CRA or is due to the hematite surfaces present in the GWS. Evaluation of this phenomenon was beyond the scope of this research.

Conclusions

The hypothesis of this research was that sorption of NISs with subsurface aquifer media is similar to sorption of NOCs and thus is a function of the hydrophobicity of the NIS. Three research objectives were addressed in the evaluation of this hypothesis: (1) to evaluate the batch sorption of NOCs and NISs with varying levels of hydrophobicity on two subsurface media; (2) to evaluate the chemical and media properties affecting the sorption of NISs in batch tests; and (3) to evaluate the sorptive migration of NISs in continuous flow column studies. The following conclusions were established as a result of this research.

(1) For the NISs (amphiphiles) evaluated in this research, hydrophobic partitioning was not adequate to account for the observed NIS sorption. It was apparent that the NISs also sorbed on the mineral surfaces of the media. The sorption of NOCs agreed with predictions based on hydrophobic partitioning.

(2) Decreasing sorption of NISs within the CO series was observed with decreasing hydrophobicity, as expected based upon hydrophobic theory. The decreasing hydrophobicity was the result of increasing oxyethylene groups (increased polarity). The ratio of K_{oc} values observed versus estimated increased with increasing oxyethylene groups. This indicates that increased interactions between the mineral surfaces and the polar moiety of the surfactants occurred with increasing oxyethylene groups, as expected based on sorbent motivated sorption.

(3) The nature of the organic matter was observed to affect the type of isotherm (linear versus Freundlich) realized for the NISs (but not for the NOCs). It is hypothesized that the mature sandstone did not have as wide a distribution

of functional groups (e.g., carboxyl, amine and hydroxyl groups) as did the more recent alluvium, and that this accounted for the nonlinear isotherms for the NISs with the alluvium. Since the NOCs would not be expected to interact as significantly with these polar functional groups, it is reasonable that the NOC isotherms would be linear with the alluvium, as observed.

(4) The sorptive transport of NISs in column studies were consistent with chromatography theory, with ratios of retardation factors in the column studies to those observed in batch studies being consistent with previously reported values (0.44 to 0.78).

(5) It is recommended that future research further evaluate the interactions of the NISs with aquifer mineral surfaces, the factors affecting the sorption of NISs with organic matters of varying composition and maturity, and the impact of NIS molecular structure on sorption.

Literature Cited

1.   Bailey, G. W. and White, J. L.   (1970).   "Factors Influencing the Adsorption, Desorption and Movement of Pesticides in Soil." In *Residue Reviews*, vol. 32. Gunther, F. A. and Gunther, J. D., Eds.; Springer-Verlag, New York, NY, pp 29-92.

2.   Hamaker, J.W. and Thompson, J.M.   (1972).   "Adsorption", *Organic Chemicals in the Soil Environment*, vol. 6, Goring, C. A. and Hamaker, J. W., Eds.; Marcel Dekker, Inc, New York, NY, pp 49-143.

3.   Rao, P.S.C. and Davidson, J.M. (1980). "Estimation of Pesticide Retention and Transformation Parameters Required in Nonpoint Source Pollution Models", *Environmental Impact of Nonpoint Source Pollution*, Overcash, M. R. and Davidson, J. M., Eds.; Ann Arbor Science, Ann Arbor, MI, pp 23-67.

4.   Weber, W.J. (1972). *Physicochemical Processes for Water Quality Control*, John Wiley and Sons, New York, NY.

5.   Karickhoff, S.W., Brown, D.S. and Scott, T.A. *Water Research*, 1979, *13*, pp 241-248.

6.   Brown, D. S. and Flagg, E. W. *Journal of Environmental Quality*, 1981, *10(3)*, pp 382-386.

7.   Scamehorn, J. F., Schechter, R. S. and Wade, W. H. *J. Colloid Interface Sci.*, 1982, *85(2)*, 463.

8.   Narkis, N., Ben-David, B. *Water Research*, 1985, *19*, pp 815-824.

9.   Hampson, J. W. , Cornell, D. G., and Micich, T. J. *Soil Science Society of America J.*, 1986, *50*, pp 1150-1154.

10.   Rouquerol, J., Partyka, S., and Rouquerol, F. (1982). "Adsorption of Non-Ionic Surfactants on Kaolin and Quartz: Correlation with Various Parameters Including the Critical Micelle Concentration", in *Adsorption at the Gas-Solid and Liquid Solid Interface*, Rouquerol, J., and Sing, S. W., Eds.; Elsevier, Amsterdam, pp 69-74.

11. Koopal, L. K., and Bohmer, M. R. (1989). "Adsorption of Non-Ionic Surfactants on a Hydrophilic Surface", in *Proceedings of Third International Conference on Fundamentals of Adsorption*, Southofen, FRG, May 7-12, 1989.

12. Denoyel, R., Rouquerol, F., and Rouquerol, J. (1987). "Compared Behaviour of Various Silica Surfaces Towards Adsorption of Poly-oxyethylenic Surfactants from Aqueous Solutions", in *Fundamentals of Adsorption*, Liapis, A. J., Ed.; Engineering Foundation, New York, NY, pp 199-210.

13. Trogus, F. J., Schechter, R. S. and Wade, W. H. *J. of Petroleum Technology*, **1979**, *31*, 769.

14. Urano, K., Saito, M. and Murata, C. *Chemosphere*, **1984**, *13*, pp 293-300.

15. Abdul, A. S. and Gibson, T. L. *Environmental Science and Technology*, **1991**, *25(4)*, pp 665-671.

16. Liu, Z., Edwards, D. A. and Luthy, R. G. (**1991**). "Sorption of Nonionic Surfactants onto Soil", Submitted to *Water Research*.

17. Lyman, W. J. (1982). "Octanol/Water Partition Coefficient", In *Handbook of Chemical Property Estimation Methods: Environmental Behavior of Organic Compounds*, Lyman, W. J., Reehl, W. F. and Rosenblatt, D. H., Eds.; New York, NY, pp 1.1-1.54.

18. *Methods of Soil Analysis. Part 1: Chemical and Microbiological Properties*, Page, A. L., Miller, R. H. and Keeney, D. R., Eds.; American Society of Agronomy, **1986**, Madison, Wisconsin.

19. Garbarini, D. R. and Lion, L. W. *Environmental Science and Technology*, **1986**, *20(12)*, pp 1263-1269.

20. Murphy, E. M., Zachara, J. M. and Smith, S. C. *Environmental Science and Technology*, **1990**, *24(10)*, pp 1507-1516.

21. Bouchard, D. L., Wood, A. L., Campbell, M. L., Nkedi-Kizza, P., and Rao, P. S. C. *Journal of Contaminant Hydrology*, **1988**, *2*, pp 209-223.

RECEIVED January 7, 1992

Chapter 15

Field Tests of Surfactant Flooding
Mobility Control of Dense Nonaqueous-Phase Liquids

J. C. Fountain

Department of Geology, State University of New York at Buffalo, Buffalo, NY 14260

Surfactants may be used to enhance pump and treat systems by increasing the solubility of organic pollutants and by increasing the mobility of non-aqueous phase liquids. However, increasing the mobility of dense non-aqueous phase liquids creates a potential for spreading of contamination through enhanced downward flow. The risk resulting from the use of surfactants may be minimized if the reduction of interfacial tensions produced by addition of surfactants is minimized. The two field tests described in this paper utilized surfactants selected for good solubilization, without excessive reduction of interfacial tension. Results of the first test indicate that efficient extraction is possible in this manner without excessive vertical mobilization.

Surfactant flooding is a modification of pump and treat technology. Water is pumped to the surface from a contaminated aquifer and treated to remove the contaminant, as in conventional pump and treat systems. Surfactant is then added to the treated water and the solution is reinjected in separate injection wells. The aquifer is treated by pumping the surfactant solution from the injection wells to the extraction wells. The surfactant solution is continuously recirculated after being treated. The use of surfactants will result in increased mobility of non-aqueous phase liquids (NAPL) in the subsurface due to the reduction in interfacial tensions (IFTs) between water and NAPL. If the NAPL is denser than water (DNAPL), the potential exists that the added mobility may result in the DNAPL moving downward into previously uncontaminated zones. The risk resulting from enhanced DNAPL mobility is site-dependent, however, in general any risk of spreading of contamination must be minimized. In this paper two field tests of surfactant flooding are described. In both cases the vertical mobility of DNAPL was minimized by selection of surfactants that produce a minimal reduction in IFTs.

0097–6156/92/0491–0182$06.00/0

Surfactants were selected primarily for their ability to increase the solubility of the pollutants present at each individual site; since surfactants can increase the solubility of virtually all organic liquids this approach can be used in remediation of aquifers contaminated by any organic compound of limited solubility e.g., halogenated solvents, creosote, PCBs.

When NAPL is present, surfactants can also be used to enhance direct recovery of NAPL through reduction of IFTs (*1*). Reduction of IFT results in a reduction of the capillary forces that normally immobilize NAPL (*2*) and hence may allow the NAPL to be driven to extraction wells by water flooding (*3*).

The addition of surfactants to groundwater may also affect sorption distribution coefficients, the wetting characteristics of the system and the tendency for emulsion formation in the aquifer. In addition, the toxicity and biodegradability of the surfactants, as with any compound added to groundwater, must be considered.   Selection of surfactants for remediation requires consideration of these parameters, as well as the details of the site hydrogeology.

In the following sections, selection criteria used for the first field test are detailed, and the results of the test are discussed in terms of surfactant performance. A comparison is then made to a second field test that is still in its initial stages.

Selection of Surfactants

Organic liquids with low solubilities in water tend to remain as separate non-aqueous phase liquids (NAPL) in the subsurface (*4-5*). Removal of NAPL from an aquifer by conventional pump and treat operations is highly inefficient because; 1) the low solubilities of common NAPLs result in extremely low concentrations of contaminant in the groundwater, thus requiring very large volumes of water to be treated (*4-5*), and 2) high IFTs between NAPL and water result in large capillary forces that immobilize the majority of the NAPL even with the highest practical hydraulic gradients (*2*).

Surfactants have been used extensively in enhanced oil recovery to lower the IFT between crude oil (a NAPL) and water, and thus allow the oil to be mobilized by water flooding (*1,6*) . Mobilization of small ganglia of NAPL requires that the IFT be reduced by at least four orders of magnitude (*6*). However, a reduction of this magnitude may introduce a serious mobility problem if the contaminants are denser than water.

DNAPLs tend to flow downward through an aquifer under the force of gravity. The downward movement will stop only when either the entire volume of DNAPL has been immobilized as residual saturation in the pores through which it has flowed (*7*) or the DNAPL encounters a layer of sufficiently small grain size that the hydraulic head produced by the accumulation of DNAPL is insufficient to overcome the entry pressure in the underlying layer (*8*). If downward flow of DNAPL is stopped by a fine-grained layer, the DNAPL will spread laterally forming a pool. The pool may be located at the base of an aquifer or may be perched on a fine-grained layer within the aquifer (*8-9*).

Both types of DNAPL immobilization are functions of capillary forces (*7-8*). Since reduction of IFT reduces capillary forces, additional vertical mobility is inevitable when surfactants are used. Reduction of IFT lowers the residual saturation of a porous media, and hence will cause partial draining of trapped ganglia within each layer. Where DNAPL has pooled on a fine-grained layer, if the IFT is reduced sufficiently, the DNAPL may penetrate the low permeability layer on which it has accumulated, and thus enter previously clean zones below (10). The risk posed by enhanced vertical mobility depends upon the geohydrology of each specific site; however, there is a significant risk that any aquitard may contain sand lenses, fractures, root holes or other paths of preferential permeability that could provide paths for contaminant migration into underlying layers.

To minimize the potential risks of vertical mobility associated with the use of surfactants, work at SUNY at Buffalo has concentrated on developing surfactant mixtures that can efficiently remove contaminants by solubilization, with a minimal reduction of IFTs. Experiments in our laboratory determined that numerous surfactants could increase the solubility of common chlorinated contaminants without reducing the IFTs between contaminant and water by more than one order of magnitude (11-12). Interfacial tensions between a 1% aqueous solution of the various surfactants and tetrachloroethylene (PCE) were measured with a DuNuoy Ring tensiometer. Several good solubilizers were found that had IFTs of > 3 dynes/cm between 1% surfactant solutions and tetrachloroethylene. Therefore the value of 3 dynes/cm as the low IFT cut off for this study.

Solubilization. Surfactants were used primarily to increase the solubility of the contaminant. The solubilization ability of surfactants was thus the prime criterion for surfactant selection. Solubilization was measured by adding 0.5 ml of PCE to 20 ml of 1% aqueous solution of surfactant and mixing. The concentration of contaminant was measured by gas chromatography after 1 week of mixing and again after a second week. Surfactants were rejected if they failed to solubilize at least 3000 ppm of PCE at 18 °C (aqueous solubility of about 200 ppm), if they came out of solution (exceeded the cloud point with the addition of PCE) or if they formed persistent emulsions (the significance of emulsion formation is discussed in the next section). Blends of those surfactants that solubilized well were subjected to the same solubilization tests to see if synergistic mixtures could be found and to see if blending would affect the formation of emulsions.

Several surfactants that were very good solubilizers when contaminant was added in small increments, tended to form thick emulsions if the PCE was added in excess. Typically the tendency to form emulsions could be greatly reduced by blending these surfactants with other good solubilizers. For example, nonylphenol ethoxylate, which was an excellent solubilizer for PCE, formed a highly viscous, persistent emulsion when PCE was added in excess. However when blended with a nonylphenol phosphate ester, this tendency to form emulsions was greatly diminished without reducing the solubilization ability.

Porous Media Experiments. The tendency for emulsion formation was found to be a function of the quantity of contaminant added and of the method and vigor of mixing. To evaluate the tendency of emulsions to form in the subsurface, and to evaluate the effect of the emulsions on the ability of the surfactant solution to extract the PCE, porous media experiments were conducted.

Two types of experiments were conducted, one used horizontal glass tubes (typically 2 cm in diameter and 30 cm long) the other glass boxes (30 cm wide, 30 cm high and 10 cm deep). In each experiment the apparatus was packed with sand of known grain size, saturated with water and contaminated with dyed PCE. Surfactant solution was then pumped through the apparatus. The glass tube and glass box experiments are described in detail in (*12*) and (*13*) respectively. The relative ability of various surfactants to extract the PCE was determined by comparing the number of pore volumes required to remove the contamination and the concentration of PCE in the effluent. Results for the most effective surfactants are presented in (*10*).

The porous media experiments indicated that those surfactants that formed persistent emulsions did not extract as efficiently as those that did not. Although in principle, emulsions could aid in extraction if they were transported with the surfactant solution, or if they aided in solubilization by increasing surface area, no such enhancement was seen. Thus, surfactants that formed persistent emulsions were eliminated from consideration.

Results also indicated that although the best extractors were good solubilizers, there was no direct correlation between solubilization ability and extraction efficiency. Work is now underway to quantify other factors, including solubilization kinetics and DNAPL surface area (a function of interfacial tension, wetting ability and spontaneous emulsion formation) that affect extraction efficiency.

Other Criteria. Any chemical added to groundwater must have low toxicity, and, as complete recovery may generally not be anticipated, should be readily biodegradable. A 1:1 blend of nonylphenyl ethoxylate (Witconol NP-100) and a phosphate ester of nonylphenyl ethoxylate (Rexophos 25-97) was selected for the first field trial based on the solubilization, IFT and extraction experiments. This blend produced good solubilization (11,700 ppm of PCE in a 2% solution at 18 °C), did not form persistent emulsions in the porous media experiments, had an interfacial tension with PCE of 3.2 dynes/cm and was the most efficient at extraction in the porous media experiments. The nonylphenyl ethoxylates have low toxicity, but are more resistant to biodegradation than many other surfactants (*14*). As the initial field test was to be done within a steel-walled cell, and the cell was to be excavated at the conclusion of the test, surfactant would not be left in the aquifer at the conclusion of the test. It was thus decided that the relative resistance to biodegradation would be acceptable.

Normally, sorption of the surfactants must also be considered. The sands at the experimental site, (Canadian Forces Base Borden), have been shown to have negligible organic carbon (about .02%) and it was thus anticipated that sorption of surfactants would not be a significant problem. Breakthrough curves measured at the site and in column experiments in the lab, have confirmed this assumption. Thus relative sorption of surfactants was not a factor in surfactant selection at this site.

A final concern was the fact that the pH of aqueous solutions of the phosphate ester is very low (about 2.5 for the 1% solution used). Injection of a solution of this low pH was deemed inadvisable, particularly since the walls of the test cell were steel and would be attacked by an acidic solution. To alleviate this problem, the surfactant solution was titrated with NaOH to a pH of 6.5 before injection.

The 3 Meter Test Cell, Canadian Forces Base Borden

The first field test was conducted in a cell constructed at Canadian Forces Base Borden, Alliston, Ontario, Canada, by the University of Waterloo's Centre for Groundwater Research. A 3 m by 3 m cell was formed by driving sheet-piling through the surficial sand aquifer and into the underlying clay (Figure 1). The sand is about 4 m thick at the test site. A second sheet-piling wall was constructed 1 m beyond the first for secondary containment. The joints in the sheet piling walls were specially constructed and sealed with bentonite.

Two hundred and thirty one liters of PCE, dyed red with Sudan IV dye was injected below the water table in an experiment conducted by Bernie Kueper, then at the University of Waterloo. The PCE was introduced in a shallow well in the center of the cell, under constant head conditions. The cell was not disturbed for approximately two months to allow flow of the PCE to stop, then the top 1 m of sand was excavated by Kueper to study the migration path of the PCE. A bentonite layer was then laid over the remaining 3 m thickness of contaminated sand and the cell was backfilled with clean sand.

Three cores were then taken using double walled corers, with the outer casing left in place to avoid leaving a pathway for fluid migration when the cores were removed. Samples were taken from the core every 5 cm and analyzed for PCE content at the University of Waterloo.

The cell contained five injection wells and five extraction wells, each screened through the entire contaminated zone. Initially, free-phase PCE was recovered from these wells by direct pumping using peristaltic pumps and small diameter plastic tubing that was lowered to the bottom of each well. Pumping of free phase continued periodically until no significant volume of free phase liquid was being recovered. Water flooding was then initiated. Water was pumped into the injection wells and out of the extraction wells, a hydraulic gradient of .094 was maintained. The extracted solution was air stripped to remove PCE. Small diameter stainless steel tubing, of sufficient length to reach the bottom of each well, was inserted into each of the wells and was pumped intermittently throughout the water flooding to remove free product. A total of 60 liters of PCE was recovered as free product (Table I).

After significant volumes of PCE were no longer being recovered by water flooding, surfactant flooding was begun. Although the system had been designed for summer use, actual pumping began on November 4, 1990. The temperature was near freezing; at this temperature, the phosphate ester was very viscous and proved difficult to mix with the available mixing equipment. The original plan was to use a 5% solution (2.5 % by weight of each of the two surfactants) but a 2% solution was used for easier mixing.

Table I. Volumes of PCE Recovered by Direct Pumping,
Water Flooding and Solubilization

Method	Volume PCE Recovered
Direct Pumping	47 liters
Water Flooding	12 liters
Solubilization (after 16 pore vols.)	79 liters
Excavation of the top 1 m (B. Kueper, unpub. data)	52 liters
Total Removed	190 liters

Results

Concentrations of PCE in the effluent were monitored daily both in the bulk effluent and in the 5 multi-level monitoring wells near the extraction wells (each had 6 sample points above the aquitard). The concentration of the effluent rose to about 4000 ppm (Figure   2), over 20 times the aqueous solubility, within one pore volume (the system had a pore volume of about 2400 gallons). Breakthrough of the surfactant in the multilevel wells occurred at the same volume as breakthrough of bromide tracer, the lack of retardation confirmed the assumption that sorption was not significant.

After seven pore volumes had been circulated, a core was taken, immediately adjacent to one of the cores taken prior to the start of surfactant flooding. Analyses of the core samples, done at the University of Waterloo, confirmed what analyses of samples from the multi-level wells had indicated:

1) PCE was rapidly being removed. The magnitude of PCE saturation was decreasing in all samples, with a significant length of the core being reduced to values less than 1% (Fountain et al; in preparation).

2) The PCE was not dropping to the base of the cell. Before surfactants were added there were three elevations at which PCE residual saturations were high. Residual saturations at all three elevations remained higher than those of other depths, although absolute values were considerably reduced after 7 pore volumes.

3) The large pool of PCE at the base of the cell was being reduced rapidly, but it, and the perched layers at higher elevations, were still present.

After 14 pore volumes were circulated another core was taken immediately adjacent to the two just discussed. Although analyses of samples from these cores have not yet been completed, photographs of the cores show that no visible PCE remains except for a narrow band, a few cm thick, at the base of the aquifer (the

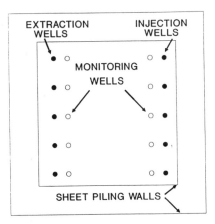

Figure   1. A plan view of the three meter cell at Canadian Forces Base Borden, Alliston, Ontario. The inner sheet piling walls form a cell approximately 3 meters on a side. Each monitoring well is a multi-level monitoring well with 12 sampling points.

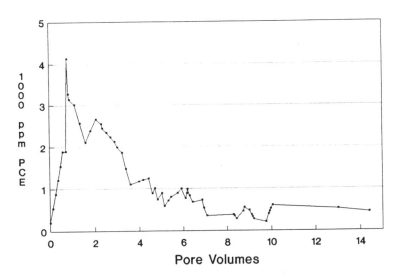

Figure   2. Effluent concentration from the surfactant flooding field test at CFB Borden. The abrupt increases in PCE concentration of the effluent correspond with times the system was restarted after shut downs. The increase at 10 pore volumes corresponded with a decrease in pumping rates from 2000 liters per day to 400 liters per day.

dye added to the PCE enables visible detection of PCE at concentrations of 1000 ppm or more).

The total volume of PCE removed after 16 pore volumes has been circulated is approximately 190 liters.   This leaves approximately 40 liters unaccounted for of the 231 liters originally added.   Estimation of the total volume of PCE remaining in the cell from the saturation of the most recent core is less than 10 liters.   The discrepancy between volume added and the volume removed will be resolved by excavating the cell later this year.   Possible explanations include volatilization of PCE from the surface of the cell, which was flooded for several months, trapping of PCE near the walls of the cell where there is little circulation or in fractures in the aquitard created by driving in the sheet piling wall.

Air Stripper Performance. Removal of the PCE by air stripping has not been as effective as desired.   Effluent concentrations from the second stage of the two-stage air stripper have ranged from a few ppm to several hundred ppm.   Air stripper performance is a function of the Henry's Law constant for a compound. Since surfactants increase the solubility of PCE and hence lower the Henry's Law constant, they will make air stripping more difficult.   However, lab scale air stripper columns have performed far better than the air strippers at the test site. The relatively poor stripper performance is due to a combination of factors including inefficient packing, low ambient temperatures, lower than optimum liquid flow rates and inefficient liquid distributor design.   Air stripper performance is beyond the scope of this paper, however, it is worth noting that the tendency of the surfactant solution to foam at high air flow velocities makes design non-trivial.   Work on air stripper design is continuing.

Mobility of DNAPL During Surfactant Flooding

One of the prime objectives of the field test was to determine the extent of induced vertical mobility of DNAPL resulting from the addition of surfactants. Three lines of evidence indicate that only minor downward mobilization occurred: 1) PCE concentrations in the individual multi-level monitoring wells rose to very high values (over 10,000 ppm) from the same elevations that showed high residual saturations in the initial three core samples. These layers remained very high for extended periods of time (the two principal lenses persisted for over 7 pore volumes), indicating that the perched lenses of DNAPL did not drop to lower levels. 2) Underlying layers, which were originally quite low in PCE concentrations, remained low, indicating that PCE was not raining down into underlying layers. 3) The core removed after seven pore volumes had been circulated, contained elevated levels of PCE in the same layers as seen in the three cores taken before surfactants were added. Although the level of residual saturation decreased, and the thickness of each contaminated layer declined, the persistence of the suspended lenses again indicates that the DNAPL lenses did not migrate downward after surfactant addition.

Although these data do not prove that vertical mobilization did not occur (indeed as stated earlier it is inevitable that some vertical movement will occur), they do suggest that large scale vertical movement was not triggered by surfactant injection. Since the aquifer is composed of well-sorted sands the data suggest that it is not necessary to have a perfect clay layer to limit vertical DNAPL movement due to surfactant addition if IFT is not reduced more than it was in this experiment (to about 3 dynes/cm).

The Corpus Christi Field Test

A second field test is now in progress in Corpus Christi, Texas, at a chlorocarbons manufacturing facility contaminated with carbon tetrachloride. Although the test area is underlain by a thick clay unit, and thus risk of vertical mobilization of the contaminant is minimal, it was decided to use a surfactant (Witconol 2722, a polysorbate) that has an IFT with carbon tetrachloride three times that of the surfactant used at the Borden site (10 dynes/cm) to see if the potential for vertical mobility could be further reduced. The polysorbate is not the most efficient solubilizer of carbon tetrachloride, but was selected as it has the highest interfacial tension of any good solubilizer. In this case, extraction efficiency was traded for a smaller reduction in interfacial tension.

In addition, since there are no containment walls at this site, the toxicity and biodegradability of the surfactants are more critical. The polysorbate is a food-grade product that is readily biodegradable.

Surfactant sorption must also be considered at the Corpus Christi site. Organic carbon is slightly higher at this site (average of 250 mg/kg), although still low. The retardation factor for a 0.5% solution of the polysorbate, determined in 5 cm diameter soil columns, is 3.5.

Analyses of cores from the Corpus Christi site yielded carbon tetrachloride concentrations of over 1100 ppm, greater than its aqueous solubility of 785 ppm. These analyses indicate that DNAPL is present in the upper 1 m of the 4 m thick aquifer. Although the field test is in its initial stages and data is as of yet insufficient for meaningful evaluation, the surfactant selection process illustrates that the surfactant solution may be tailored to site-specific considerations.

Conclusions

1. Surfactants have been shown to be an effective method of increasing the efficiency of pump and treat operations. Maximum effluent concentrations of over 4000 ppm were more than 20 times the aqueous solubility of PCE and hence PCE was being removed at least 20 times as efficiently as it could have been by water pumping. The effluent remained above the aqueous solubility throughout the test compared to concentrations of 10 ppm or less that are typical for extended pump and treat operations.

2. The decrease in PCE saturation observed between the start of surfactant flooding and after seven pore volumes indicate that surfactant flooding was rapidly removing the DNAPL from the aquifer. Removal of DNAPL from an aquifer has been shown to be the principal difficulty facing pump and treat remediation operations (4,16).

3. The use of surfactants need not result in excessive vertical mobility if the surfactants used do not reduce interfacial tensions below a few dynes/cm.

Acknowledgments

The Canadian Forces Base Borden Site is operated by the University of Waterloo, who provided all physical facilities at the site including the test cell, monitoring and pumping wells and all coring. Bob Starr of the Univeristy of Waterloo provided invaluable help throughout the project. Discussions with Bernie Keuper, Stan Feenstra and John Cherry also contributed significantly. Thesis research by Mike Beikirch, Tom Middleton and Andrew Klimek provided the background for this work and Middleton and Beikirch were the primary operators of the Borden test.

This research was sponsored by the New York State Center for Hazardous Waste Management, the University of Waterloo's Solvents in Groundwater Program and the E.I. Du Pont Corp.

Literature Cited

1. Vigon, B.W.; Rubin A.J. *J.Water Pollution Contr. Fedr.* **1989**,v.61, pp.1233.
2. Wilson, J.L.; Conrad, S.H. *Proc. NWWA/API Conf. Petrol. Hydroc.;* Water Well J. Pub. Co., Worthington, OH, **1984**,pp.264-298.
3. Sale, T.; Piontek, K.; Pitts, M. *Proc. NWWA/API Conf. Petrol. Hyroc.;* Water Well Pub. Co., Worthington, OH, **1989**, pp.487-503.
4. Mackay, D.M.; Cherry, J.A. *Environ. Sci. Tech.* **1989**, v. 23, p. 630.
5. Mercer, J.W.; Cohen, R.M. *J. Contam. Hydrology*, **1990**, v. 6, pp. 107.
6. Miller, C.A.; Qutubuddin, S. *In: Interfacial Phenomenon in Apolar Media;* Eicke, H.F; Parfitt, C.D., Eds. **1987**; Marcel Dekker, New York, pp. 117-185.
7. Hunt, J.R.; Sitar, N.;Udell, K.S. *Water Resour. Res.* **1988**, v. 24, p. 1247.
8. Kueper, B.H.; Frind, E.O. *Water Resour. Res.* **1991**, v. 27, pp. 1049.
9. Scwille, F. *Dense chlorinated solvents in porous and fractured media.* Pankow, J.F., Translator; Lewis Pub., Chelsea, MI,**1988**, 145 pp.
10. Fountain, J.C.; Klimek, A.; Middleton, T.; Beikirch T. *J, Hazard. Mater.*, **1991**, v. 28, pp. 295.
11. Fountain, J.C.; Klimek, A.; Beikirch, M.; Middelton, T.; Hodge, D. *Proc. Aquifer Reclamation and Source Contl. Conf.;* New Jersey Inst. Tech.,Newark, NJ, **1990**, p.1.
12. Klimek, A. *In-Situ Chemically Enhanced Dissolution of Chlorinated Hydrocarbons*; M.A. Thesis, Geology, Univ. Buffalo, Buffalo, NY. **1990**.
13. Beikirch, M. *Experimental Evaluation of Surfactant Flushing for Aquifer Remediation;* M.A. Thesis, Geology, Univ. Buffalo, Buffalo, NY.**1991**.
14. Schick, M.J. *In: Nonionic Surfactants,* Schick, M.J., Ed. Surfactant Science Series; Marcel Dekker, New York, **1967**, v.1, pp.971.
15. Mackay, R.A. *In: Nonionic Surfactants Physical Chemistry*, M.J. Schick, Ed., Surfactant Science Series, Marcel Dekker, New York, **1987**, v.23., p.297.

RECEIVED December 11, 1991

ORGANIC COMPOUNDS

Chapter 16

Landfill Leachate Effects on Transport of Organic Substances in Aquifer Materials

F. M. Pfeffer and C. G. Enfield

Robert S. Kerr Environmental Research Laboratory, U.S. Environmental Protection Agency, Ada, OK 74820

The effect of dissolved organic carbon (DOC) in landfill leachate on the transport of a hydrophobic organic compound through saturated aquifer material was investigated. Leachate DOC was found to be complex; attempts to characterize the organic matrix were not successful. Two hydrophobic compounds of environmental significance were evaluated for 1) their partitioning to DOC in leachate, and 2) the effects of leachate on their partitioning to aquifer material. Results showed that the leachate reacts with aquifer material to increase partitioning to the stationary phase and that the DOC in the landfill leachate enhances mobility of the hydrophobic compounds. The conclusion was that at present, we do not know how to predict the impact of a given leachate on a given aquifer material without experimental measurements.

The transport of organic chemicals through saturated aquifer material is governed by many physical, chemical, and biological processes. The degree of organic chemical sorption, and therefore the retardation relative to the movement of water, is a function of solute hydrophobicity and the amount of sorbent hydrophobic phase. Properties that reflect chemical hydrophobicity, such as aqueous solubility (S), octanol:water partition coefficient (K_{ow}), and reverse-phase chromatographic retention time, have been successfully used to predict the magnitude of sorption *(1-6)*. In addition, the importance of soil organic carbon in hydrophobic sorption has been repeatedly demonstrated *(7,8)*. Enfield et al. *(9)* questioned the assumption that chemicals in solution act independently of each other and demonstrated that dissolved organic carbon (DOC) in the fluid phase could increase the mobility of hydrophobic compounds, while Bouchard et al. *(10)* and Lee et al. *(11)* questioned the assumption of a stationary and unchanging hydrophobic phase and demonstrated that cationic surfactants could significantly increase the partitioning of neutral organics in an aquifer material.

Landfill leachate is a complicated mixture of organic chemicals which has been extremely difficult to characterize *(12-14)*. We are proposing three conceptual mechanisms for the DOC in landfill leachate to affect the transport of hydrophobic compounds. First, the DOC in the leachate may increase the solubility of hydrophobic chemicals in the mobile phase where the DOC might behave like a cosolvent or may act like a surfactant partitioning hydrophobic compounds into a mobile organic phase. Second, the leachate may extract a portion of the soil's organic carbon, reducing the partitioning compared to that observed without leachate. Third, if polar or ionized organic material is in the leachate, the partition coefficient to the stationary (soil) phase may increase by replacement of inorganic ions with polar or ionizable organic compounds.

To evaluate the significance of these potential mechanisms, three independent types of analyses were performed. First, a vapor phase transfer study was performed demonstrating the ability of the leachate to concentrate hydrophobic chemicals suggesting an increased potential for contaminant mobility. Second, batch sorption isotherms were measured. The soil in the batch studies was pretreated with either $CaCl_2$ or leachate. This was for the purpose of determining if there was an interaction between chemicals in the leachate and the soil. Finally, in anticipation of future studies, an attempt was made to produce a chemical "fingerprint" of the organic matrix in the leachate. This was to determine if changes in the matrix were occurring during the study.

Experimental Section

Materials and Instrumentation.   The hydrophobic compounds studied were hexachlorobenzene (HCB) and benzo(a)pyrene (BaP). For HCB and BaP, the aqueous solubilities (S) are 6.0×10^{-3} mg/l and 3.8×10^{-3} mg/l, respectively, and the octanol:water partition coefficients (K_{ow}) are 2.60×10^6 and 1.15×10^6, respectively *(15)*. Because of the low solubilities of these study compounds it was necessary for purposes of quantitation in partitioning experiments to use ^{14}C-labeled compounds obtained from Sigma. The activities of the compounds were: HCB=10.76 mCi/mmol; BaP=5.0 mCi/mmol. Each ^{14}C-labeled compound was initially dissolved in a very small amount of hexane and then diluted with methanol to make a spiking solution for the aqueous experiments. Three leachates were used in the study: one from Denmark collected in 1987 and two from Oklahoma. The Oklahoma samples were collected in 1985 and 1987 in Norman, Oklahoma from a well located in a closed municipal landfill. The Denmark leachate was preserved with mercuric chloride (0.2 g $HgCl_2$/l) before shipment. All leachates were stored at 4°C in glass containers with minimum headspace. Leachate total organic carbon (TOC), reported as mgC/l, was measured with a Dohrman DC-80 analyzer. The leachate TOC values were: Denmark (355); Norman 1985 (200); Norman 1987 (142). The aquifer material was collected at a depth of 5 meters from a location near Lula, Oklahoma. The material was air dried and sieved to < 2mm. The TOC of this material was 0.36 ± 0.08 g/kg (12 replicates) as measured with a Leco WR-12 analyzer. HCB and BaP were quantitated with a Beckman LS7800 liquid scintillation counter using 1 ml aqueous sample and 6 ml Beckman HP cocktail. Extractable organics were identified by GC/MS using a Finnigan 4600 system. Purgeable organics were identified by GC/MS using a Tekmar purge and trap system and a Finnigan 4500 system. Pyrolysis-GC/MS was performed using a CDS Pyroprobe 120 and a Finnigan Ion Trap Detector 700. Infrared spectra were obtained with a Bio-Rad FTS-45 using a diffuse reflectance

(DRIFT) accessory. A metals scan was performed using a Thermo Jarrell-Ash 975 ICP spectrometer.

Procedures. Experiments were conducted to measure the equilibrium partition coefficient K_p, defined as C_o/C_a, where C_o is the concentration of the study compound on DOC in the leachate and C_a is the corresponding aqueous concentration. A vapor phase transfer apparatus shown in Figure 1 was used, consisting of boiling flasks connected by an adapter. Leachate (50 ml) was placed in one flask and deionized water (50 ml) was placed in the other flask. Each flask was spiked at <50% of its water solubility with either ^{14}C-labeled HCB or BaP. Methanol concentration after spiking was ≤0.09% (by volume) in all instances. The flasks were then monitored for about three weeks as a function of time by withdrawing 1 ml samples from both flasks for scintillation counting for 60 min., or until the 95% confidence limit of ±1.40% was obtained, whichever occurred first. All experiments were run in duplicate, and the results were averaged for presentation. Aquifer material was not used in these studies.

Batch sorption studies were performed to: (1) measure the equilibrium partition coefficient K_d, defined as C_s/C_a, where C_s is the concentration of study compound on the solids; and (2) determine the impact of sample handling on the measured K_d. Partition coefficients were measured using 2 g of aquifer material and 45 ml of liquid phase in a 50 ml glass culture tube with Teflon cap liner. Aquifer material was washed or pretreated with either $CaCl_2$ or leachate sequentially 10 times prior to conducting HCB isotherms in the presence of either $CaCl_2$ or leachate as supporting electrolyte. After each washing the slurry was centrifuged for 3 hrs at 1100 g and the equilibrating solution was decanted and replaced with "fresh" equilibrating solution. When prewashing was completed, the volume was corrected by weight such that the soil:solution ratio was known. Four experiments were evaluated for each leachate: a $CaCl_2$ wash and the isotherm conducted with $CaCl_2$ (designated CC-isotherm); a $CaCl_2$ wash and the isotherm conducted with leachate (CL-isotherm); a leachate wash and the isotherm conducted with $CaCl_2$ (LC-isotherm); and a leachate wash and the isotherm conducted with leachate (LL-isotherm). The ^{14}C-labeled HCB was added in four concentrations to two replicates just prior to the sorption experiments. The samples were slowly rotated end over end for 24 hrs to establish equilibrium. Blanks (no soil) were used to correct for system losses. Direct measurement of the concentration of the study compounds on the solids was attempted by sampling the centrifuged solids and correcting for water content. The precision of this technique was found to be unacceptable. (For concentrations in the aqueous phase ranging from 10-40% of the solubility limit, the standard deviation for the mass on the solid phase was ± 38-62% based upon 4 sets of 8 replicates each). Therefore C_s was calculated by difference based upon the measured C_a and the total added in the sorption experiment corrected for losses.

An attempt was made to obtain a "fingerprint" characterization of the DOC. For this purpose the Norman, Oklahoma landfill was again sampled in 1989 from the same well. The sample was filtered using an all glass apparatus and a 0.45 μm Millipore HA filter with a Millipore AP prefilter. Filtrates were stored at 4° C in glass containers with minimum headspace. For purposes of FTIR and pyrolysis/GC/MS analyses, a 40 ml subsample was frozen in a glass container and lyophilized to dryness using a Labconco freeze dry system. A subsample was prepared in KBr at a concentration of 1 wt% prior to analysis by FTIR. A 100 µg subsample was placed in a quartz tube for pyrolysis in the coil probe.

Results and Discussion

Vapor Phase Transfer Experiments. The results of three experiments to determine K_p using the vapor phase transfer apparatus are shown in Figure 2. The values for log K_p estimated from the data are 4.8-5.0 for BaP and 4.3 for HCB. This assumes that the carbon in the leachate behaves as a second dispersed phase rather than as a cosolvent. The time required to approach equilibrium for HCB was much shorter than that for the BaP which can be expected based on their vapor pressures. Vapor pressures are: HCB= 1.09×10^{-5} mm Hg @ 20° C; BaP = 5.6×10^{-9} mm Hg @ 25° C (15). The partition coefficients for the BaP with the Denmark and Norman 1985 leachates are similar.

According to Hutchins et al. *(16)*, Carter developed a humic materials:water partition coefficient versus an octanol:water partition coefficient relationship as

$$\log K_p = 0.709 \log K_{ow} + 0.748$$

West *(17)* measured partitioning of hydrophobic materials to ground water humic materials and developed the following correlation between water solubility and humic material:water partition coefficient:

$$\log K_p = -0.923 \log S \text{ (mg/l)} + 3.294$$

A comparison between these two regression models yield log K_p's for HCB of 5.3 and 5.3, respectively, and log K_p's for BaP of 5.0 and 5.5, respectively. Carter's relationship gives a slightly lower estimate of K_p for BaP. The values obtained experimentally are consistent with the estimates from both regression models. These experiments demonstrate the potential for increased mobility by leachate compared to water but are not conclusive in determining if this is a phase transfer phenomenon or a cosolvency phenomenon.

Batch Sorption Studies. When soils are dried, the native organic carbon is oxidized and a portion becomes water soluble. During batch sorption experiments this water soluble carbon can interfere with the interpretation of experimental data. When the impact of the dissolved phase organic carbon is the subject of the study, it is necessary to minimize the effect of the native organic carbon which has come into solution. Washing the soil prior to experimentation was chosen as acceptable for removing native soluble organic carbon. Experimental measurements (not shown) revealed that ten washings removed ≤2% of the solids TOC (as measured in the aqueous extracts) and that new organic carbon was not solubilized with additional washing. The effect of landfill leachate on partitioning is shown in Figures 3 and 4. In each of these figures the curve labelled CC is a reference curve. The reference curve was obtained by washing the soil with 0.01 M $CaCl_2$ and measuring the HCB partitioning to the Lula soil in 0.01 M $CaCl_2$. The data shown in Figure 3 was collected to see if contacting the soil with leachate prior to measuring a sorption isotherm could measurably change the properties of the soil. To do this, rather than washing the soil with 0.01 M $CaCl_2$ the soil was washed with the different leachates. The sorption isotherm was then measured as before in $CaCl_2$. The Denmark leachate appears to increase the partitioning of HCB to the Lula soil. This would reduce the mobility of HCB. However, both Norman leachates decreased the

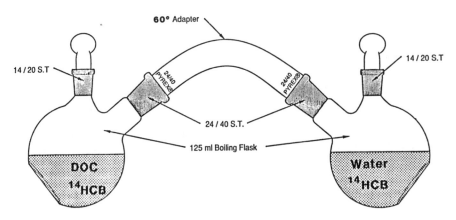

Figure 1.   Vapor phase transfer apparatus

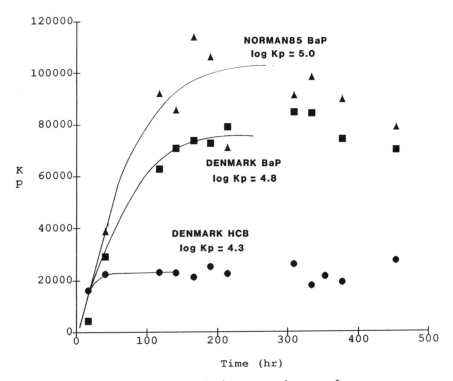

Figure 2.   Kp results from vapor phase transfer

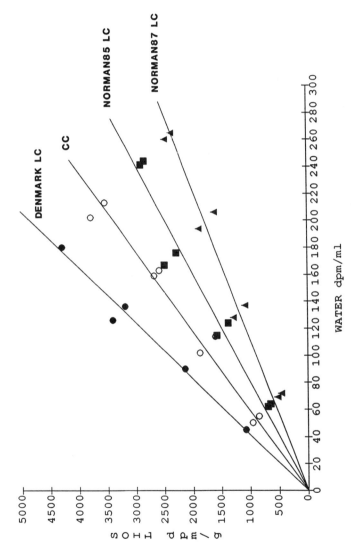

Figure 3. Effects of prewashing with leachate

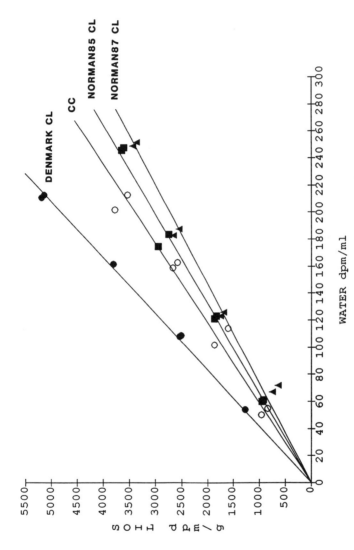

Figure 4.   Effects of leachate in the experiment

partitioning of HCB to the Lula soil. It appears that some component of the Norman leachates was able to extract a significant portion of the native organic carbon thereby increasing the potential mobility of HCB. No attempt was made to determine the stability of the system or if soils had reached quasi equilibrium. The partition coefficients are all within a factor of 3. If the soil/leachate systems were at equilibrium, the environmental significance of the leachate on partitioning is measurable but not dramatic. The same pattern but with a smaller shift is evident in data generated when the soil was washed with $CaCl_2$ and leachate was introduced during the isotherm measurements (Figure 4). The same influence of the Norman leachates on native organic carbon is apparent. It appears that the impact of the leachate on the soil is more significant than the ability of the leachate carbon to increase pollutant transport by acting as a cosolvent or a separate phase.

Leachate Characterization. The TOC of the Norman leachate sampled in 1989 for characterization purposes measured 88 mg/l C. Volatile and nonvolatile organics identified by GC/MS accounted for <1% of the leachate TOC. A metals scan showed that the leachate was essentially void of heavy metals but high in Ca, Mg, Na, and K.

The results of pyrolysis/GC/MS for the freeze dried leachate are found in Figure 5. Compound peaks appear small against a noisy background. One explanation for this is that the predominant materials in the solid sample are salts. This could present a problem of salt-organic interaction occurring at the temperature of pyrolysis similar to what has been reported by Alcaniz et al. *(18)*. The source of the siloxanes was found to be the freeze dry apparatus. The profile of remaining organics identified are consistent with that reported for humic and fulvic acids *(19)*. Any further work with pyrolysis/GC/MS would necessitate preliminary removal of the inorganic matrix.

An FTIR spectrum of the freeze dried leachate obtained with the DRIFT accessory is found in Figure 6. Spectra for $CaCO_3$ and Na_2SO_4 are also displayed to demonstrate that much of the peak profile in the region of 1500-600 wavenumbers could be accounted for by the presence of these two salts. One region of particular interest for organics characterization is 2960-2930 wavenumbers. This region is void of distinct peaks, most probably due to the very dilute organic matrix in the freeze dried residue. Again, further effort with FTIR would require preliminary preparation to provide a residue which is much more concentrated in organic matter.

Conclusions

It was shown that the carbon in landfill leachates has the potential for acting as either a cosolvent or a surfactant. If it is assumed that the carbon will behave as a surfactant, the partition coefficients observed are similar to what others have reported for humic materials. The presence of leachate also modified the partition coefficient to soil. Some leachates reduced the soil's ability to partition HCB while another leachate increased the soil's ability to partition HCB. This reproducible but inconsistent effect of leachate on the soil makes it currently impossible to predict the impact of leachate on chemical transport without experimental measurements. It was not possible to quantify the chemical components in the leachate which caused these observations. Leachate can impact chemical transport; however, the magnitude of this impact was less than a factor of 3 for the soil and leachates studied.

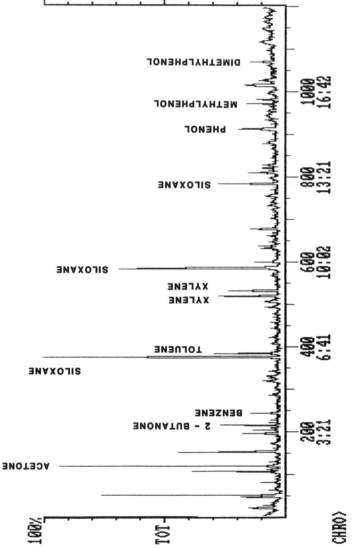

Figure 5. Pyrolysis/GC/MS results

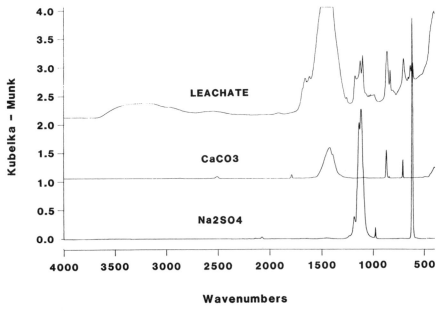

Figure 6.   FTIR spectra (DRIFT accessory)

Acknowledgements

The aid of Vallerie Boyd in conducting the bench experiments is gratefully acknowledged. We thank Garmon Smith and Dennis Fine for performing all of the GC/MS analyses, and Don Clark for performing the metals scan. We acknowledge Neil Adrien of Oklahoma University, Norman, Oklahoma and Thomas Christensen, Technical University of Denmark, Lyngby, Denmark for providing landfill leachates.

Although the research described in this article has been funded wholly by the U.S. Environmental Protection Agency, it has not been subjected to Agency review and therefore does not necessarily reflect the views of the Agency and no official endorsement should be inferred. This project was conducted under an approved Quality Assurance Project Plan and the procedures therein specified were used without exception. Information on the plan and documentation of the quality assurance activities and results are available from the Principal Investigator.

Literature Cited

(1)   Carlson, R.M.; Carlson, R.E.; Kopperman, H.L. *J. Chromatogr.* **1975**, *107*: 219-223.
(2)   Kenaga, E.E.; Goring, C.A.I. *Aquatic Toxicology*; Am. Soc. Testing and Materials. **1980**; Spec. Tech. Pub. No 707, pp. 78-115.
(3)   Karickhoff, S.W. *Chemos.* **1981**, *10*: 833-846.
(4)   Brown, D.S.; Flagg, E.W. *J. Env. Qual.* **1981**, *10*: 382-386.
(5)   Weber, W.J.; Chin, Y.; Rice, C.P. *Wat. Res.* **1986**, *20*: 1433-1442.
(6)   Chin, Y.; Weber, W.J.; Voice, T.C. *Wat. Res.* **1986**, *20*: 1443-1450.
(7)   Karickhoff, S.W.; Brown, D.S.; Scott, T.A. *Wat. Res.* **1979**, *13*: 241-248.
(8)   Chiou, C.T.; Porter, P.E.; Schmedding, D.W. *Env. Sci. Tech.* **1983**, *17*: 227-231.
(9)   Enfield, C.G.; Bengtsson, G.; Lindqvist, R. *Env. Sci. Tech.* **1989**, *23*: 1278-1286.
(10)   Bouchard, D.C.; Clark, D.A.; Powell, R.M. *J. Env. Sci. Health.* **1988**, *A23(6)*: 585-601.
(11)   Lee, I.F.; Crum, J.C.; Boyd, S.A. *Env. Sci. Tech.* **1989**, *23*: 1365-1372.
(12)   Chian, E.S.K.; DeWalle, F.B. *Env. Sci. Tech.* **1977**, *11*: 158-163.
(13)   Harmsen, J. *Wat. Res.* **1983**, *17*: 699-705.
(14)   Weis, M.; Abbt-Braun, G.; Frimmel, F.H. *The Sci. of the Tot. Env.* **1989**, *81/82*: 343-352.
(15)   Mabey, W.R.; Smith, J.H.; Podoll, R.T.; Johnson, H.L.; Mill, T.; Chou, T.W.; Gates, J.; Patridge, I.W.; Jaber, H.; Vandenberg, D. *Aquatic Fate Process Data for Organic Priority Pollutants*; Prepared by SRI International for Monitoring and Data Support Division, Office of Water Regulations and Standards, U.S. Environmental Protection Agency, Washington, D.C. **1982**.
(16)   Hutchins, S.R.; Tomson, M.B.; Bedient, P.B.; Ward, C.H. *CRC Critical Reviews in Env. Control.* **1985**, *15(4)*: 355-416.
(17)   West, C.C. Ph.D. Thesis, Rice University, **1984**.
(18)   Alcañiz, J.; Romera, J.; Comellas, L.; Munne, R.; Puigbo, A. *Sci. Tot. Env.* **1989**, *81/82*: 81-99.
(19)   Alcañiz-Baldellou, J.M.; Andrés-Canadell, *J. Anal. Appl. Pyrol.* **1982**, *4*: 241-256.

RECEIVED December 18, 1991

Chapter 17

Clay and Immiscible Organic Liquids
Greater Capillary Trapping of the Organic Phase

Robert E. Mace[1] and John L. Wilson

Department of Geoscience and Geophysical Research Center, New Mexico Institute of Mining and Technology, Socorro, NM 87801

Clay can have a profound effect on the entrapment of immiscible organic liquids in the saturated zone of groundwater aquifers. Column studies of a clean, uniform sand with essentially no clay and a high ionic strength aqueous solution show that trapped organic liquid occupies 16.9% of the pore space. This percentage is consistent with more idealized studies using glass beads. Adding clay at 3% of the total dry material weight increases the amount of trapped organic to 30.3%, a considerable difference in entrapment for a small change in texture. Styrene polymerization experiments show explicitly that a sand with clay has a much higher organic liquid entrapment than the same sand cleaned of clay. Experiments in etched glass-plate micromodels with the same solution chemistry and clay show flocculated clays blocking and constricting pore throats. It appears that clay particles, originally attached to the pore walls, detach and are mobilized by the liquid-liquid interface. The interface transports and actively compresses the mobilized clay particles into pore throats. The compressed clays yield pores more resistant to fluid flow, leading to by-passing. Therefore, it appears that organic liquids play an active role in moving and rearranging colloids, with a concomitant increase of capillary trapping.

Clays have important effects on aquifer material properties. It is well known, for instance, that clay migration can lead to a decrease in formation permeability (*1,2,3,4*). This phenomenon is attributed to particles becoming trapped by filtering in the porous media and reducing the available paths for water flow. Petroleum engineers refer to this problem as formation damage (*5*). There has been much less work addressing how clays affect the capillary trapping of immiscible, non-wetting fluids (*6*). This issue is of concern in aquifer remediation of non-aqueous phase liquids (NAPLs) because clays may cause a considerable increase in the NAPL trapping efficiency of aquifer material and possibly impede existing remediation techniques.

Wilson et al. (*7*) studied the effect of different fluid properties on the capillary trapping of NAPLs in unconsolidated, water-wet porous media initially saturated with

---

[1]Current address: Bureau of Economic Geology, The University of Texas at Austin, Austin, TX 78712

0097–6156/92/0491–0205$06.00/0

water. They found that, at low capillary and Bond numbers, interfacial tension, NAPL viscosity, and NAPL density had little effect on how much NAPL was trapped in the porous media. They also looked at various sands of similar composition but slightly different texture: a beach sand, a fluvial sand, and an aeolian dune sand. Theoretically, a pack of uniform sand should yield nearly the same NAPL residual saturation as a glass bead pack due to similar geometry and capillary forces (7,8). Morrow and Chatzis (9), Chatzis et al. (10), and Morrow et al. (11) found residual NAPL saturations in water-wet, uniformly sized glass beads to be 14 to 16% of the pore space. Using the same columns used in this study, Wei (12) obtained results of residual saturations in glass bead packs in the range of 17-18%. The difference between Wei's and the literature's saturations was attributed to column design, which may have allowed an additional amount of NAPL to be by-passed at the bottom of the porous media pack. Wilson et al. found the beach sand and fluvial sand to behave as expected with NAPL residuals of 17.6±1.5 % and 15.8±1.8 %, respectively. However, the 'Sevilleta' sample of aeolian sand had a much greater residual saturation than expected (27.1±1.7 %). Wilson et al. noticed that the Sevilleta sand contained a small amount of clay which was evident upon packing (about 2% clay and silt by weight), while the others did not. They hypothesized that the reason the Sevilleta sand had greater NAPL trapping than the other sands was due to the clay content. The objective of this study was to investigate this hypothesis and to quantify the role of the clay fraction on the trapping of NAPLs.

Background

In porous media, a NAPL is under the influence of three forces: gravity (or buoyancy) forces, viscous forces, and capillary forces. In typical porous media under ambient aquifer conditions, NAPLs are predominantly affected by capillary forces, at least in fine grained materials. The Young-Laplace equation of capillarity can be defined as follows, assuming a spherical menicus and cylindrical capillary tube:

$$P_c = P_{nw} - P_w = \sigma C = \frac{2\sigma}{r_c} = \frac{2\sigma \cos\theta}{r_t}$$

$$
\begin{aligned}
P_c &= \text{capillary pressure} \\
P_{nw} &= \text{pressure in the non-wetting phase (NAPL)} \\
P_w &= \text{pressure in the wetting phase (water)} \\
\sigma &= \text{fluid-fluid interfacial tension} \\
C &= \text{curvature of the fluid-fluid interface} \\
r_c &= \text{radius of curvature} \\
\theta &= \text{contact angle of the wetting fluid} \\
r_t &= \text{radius of the capillary tube}
\end{aligned}
$$

Capillarity depends on the fluid-fluid interfacial tension (σ), wettability of the fluids to the solid (θ), and the geometry of the porous media (r_t). These parameters influence the capillary trapping of NAPLs in porous media.

Snap-off. Entrapment of NAPLs can occur in saturated, water-wet porous media by two methods: snap-off and by-passing (7,10,13,14,15). Snap-off occurs when the non-wetting phase is displaced from a pore body into a pore throat by a wetting phase. Figure 1 shows snap-off occurring in a pore. The wetting phase moves along the sides of the pore walls where it displaces the non-wetting phase. As the wetting phase progresses, it reaches the end of the pore body before the non-wetting phase can exit. As the wetting phase begins to enter the pore throat, the non-wetting phase 'snaps-off' and remains in the pore as a trapped singlet. Snap-off depends on pore geometry and wettability. If the ratio between the size of the pore body and the pore throat (pore aspect ratio) is small , complete displacement may be possible. If this ratio is large, trapping of the non-wetting phase is likely. In Figure 1, the advancing fluid wets the pore walls ($\theta = 30°$). Theoretically, if the fluids were intermediate wet ($\theta = 90°$), snap-off would not occur.

By-passing. By-passing occurs when the non-wetting fluid is displaced by a wetting fluid in a complex pore structure. By-passing is best described by using the pore doublet model (*14,16*). Part A of Figure 2 shows a pore doublet in which no by-passing occurs because of similar pore geometry in both the upper and lower arms. Part B of Figure 2 shows by-passing in a pore doublet due to the different geometry in the lower and upper arms. The greater capillary forces in the upper arm lead to the by-passing of the non-wetting phase in the lower arm.. The pore doublet model demonstrates by-passing on a pore level, but by-passing occurs on larger scales. In more complicated pore structures, such as sandpacks, the by-passed non-wetting phase can be complex in shape, occupying many pore bodies (*7,17*). Also, in heterogeneous soils, large pockets of organic can be by-passed in coarse lenses due to capillary forces (*7*).

It is clear from the snap-off and by-passing phenomena that soil texture may be important in the trapping of fluids in porous media. Morrow (*18*) showed that particle and pore shape had little effect on wetting phase residual saturations. He concluded that the magnitude of the residual saturations would be dominated by heterogeneity in the pore structure. By using different sized and sorted glass beads, Chatzis et al. (*10*) found that soil texture had no significant influence on the non-wetting phase residual saturations.

Soil and Fluid Characteristics

Soil Characteristics. The porous media used in this study was Sevilleta sand from aeolian sand dunes 15 miles north of Socorro, New Mexico (*7,19*). This media was a well sorted sand of sub-angular to sub-rounded particles with a median particle diameter of 0.3 mm. The composition of the Sevilleta sand was mostly quartz ($72 \pm 5\%$ by number of grains), with smaller amounts of feldspar ($11 \pm 2\%$) and lithic fragments ($16 \pm 3\%$). The lithic fragments were smaller in size relative to the quartz and comprised roughly 5% of the sand by volume. Clay comprised approximately 0.4% by weight. Organic carbon content in the Sevilleta sand was less than 0.02%. Particle density was 2.65 g/cm^3. Hydraulic conductivity and intrinsic permeability were 1.17 x 10^{-2} cm/s and 1.15 x 10^{-8} darcys, respectively. Wilson et al. (*7*) found the Sevilleta sand to be very water wet.

To evaluate how different clay contents affected the trapping of the NAPL in the soil, it was necessary to clean the fine particles from the Sevilleta sand and artificially add clay. A decanting procedure involving a dispersant removed the clays (*19*). Unavoidably, other fines were also removed. The decanting procedure removed 2.1% of the dry sample weight of which 17.6% was determined to be clay (< 2 microns) by the hydrometer method.

The clay subsequently added to the clean Sevilleta sand was Ca-montmorillonite purchased from the Source Clays Repository (c/o Professor William D. Johns, Source Clays, Department of Geology, University of Missouri, Columbia, Missouri 65211.) This "Cheto" clay was gathered from Apache County, Arizona (*20*). Particle density of the clay was 2.07 g/cm^3. The behavior of the clay throughout the experiments indicated it was water wet.

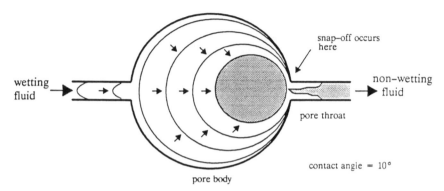

Figure 1.   Snap-off in a water wet, high aspect ratio pore (adapted from ref. 14).

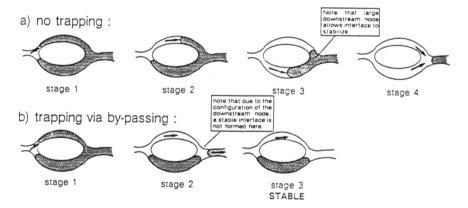

Figure 2.   By-passing shown by pore doublets (adapted from ref. 14).

The procedure used to make the sand-clay mixtures involved adding enough distilled, de-ionized water to wet the grains and then adding in a predetermined amount of clay to the sand. The sand and clay were mixed briskly and then allowed to air dry. Sand-clay mixtures comprised 0.5%, 1.0%, 2.0%, 3.0% and 5.0% clay content by total sample dry weight.

Fluid Characteristics. Three fluids were used in this investigation: an aqueous phase and two NAPL phases. The aqueous phase was a distilled, de-ionized, and de-aired aqueous solution that was 0.037 M $CaCl_2$. To de-air the water, it was boiled and kept under a vacuum. $CaCl_2$ was added to the water as a flocculating agent to prevent clay particles from migrating through the column in the initial experiments conducted by Wilson et al.(7). The pH of the aqueous solution was 5.73 and was not controlled during the course of the experiments.

The NAPL phase used in the column experiments was Soltrol-130, a mixture of C_{10} to C_{13} isoparaffins manufactured by the Phillips 66 Company. Its low solubility in water, low volatility, and low toxicity made this colorless, mild odored fluid relatively safe for laboratory column experiments. The NAPL phase used in the polymerization experiments was styrene. Styrene's immiscible behavior with respect to water, low viscosity, and ability to harden while in contact with water made this organic liquid ideal for the polymerization experiments. Unlike Soltrol-130, experiments with styrene were conducted with a respirator, gloves, and a fume hood to prevent contact with the fluid. Table I lists the different fluid properties of the phases.

Table I: Summary of Aqueous and NAPL Fluid Properties[1]

liquid	specific gravity	density (g/cm3)	kinematic viscosity (cst)	dynamic viscosity (cp)	interfacial tension with 0.3% CaCl2 solution (dynes/cm)	surface tension (dynes/cm)
aqueous phase	1.003±0.002	1.000±0.002	0.98±0.01	0.98±0.01	not applicable	72.0±0.4
Soltrol 130	0.755±0.002	0.753±0.002	1.93±0.01	1.93±0.01	47.8±1.2	19.1±0.3
styrene[2]	0.906±0.002	0.903±0.002	0.89±0.01	0.81±0.01	35.3±3.0	31.9±0.3

SOURCE: Data are from ref. 7.

[1] all measurements were taken at temperatures between 22° and 24° Celcius.

[2] with the addition of the fluorescent dye 9,10-diphenylanthracene (0.6% by weight), the interfacial tension of styrene with 0.037 M $CaCl_2$ solution was 30.9±0.3 dynes/cm.

Experimental Procedures.

Three different experiments were used to investigate how clay affected the trapping of NAPLs: short columns, styrene polymerization columns, and glass-plate micromodels.

Short Columns. Short, sand packed, glass columns were used to quantify the amount of NAPL trapped in the sand and to determine the liquid saturation-capillary pressure properties of the porous media. The column was developed by Wilson et al. (5) and consisted of a modified gas chromatographic column with two Teflon end caps. Each endcap screwed into the glass column and sealed tightly against the glass with an o-ring. The porous media chamber in the column measured 5 cm in height and 5 cm in diameter. The bottom endcap included a fritted glass plate for filter support and to facilitate uniform flow. Epoxy sealed the frit to the endcap. To allow water, but not Soltrol-130, to pass through the endcap, a water-wet Magna 66 nylon filter with a pore diameter of 0.22 micron was glued onto the glass frit. To protect the integrity of the nylon filter from soil abrasion, a paper filter was glued on top of the nylon. The top endcap was similar to the bottom. However, instead of a glass frit, there was solid Teflon with radial grooves etched into the cap to facilitate flow. A polypropylene scrim, a paper filter, and another scrim were glued to the top endcap. The paper filter prevented any fines from escaping the column and the coarse scrims provided support. Fluid volumes within the column were measured gravimetrically (7).

Sand was packed into the column under a centimeter of water to avoid layering. As soil was poured into the column, it was packed constantly to avoid the formation of any clay layers. The column was then saturated by flowing de-aired water through the column (7,19). The column was connected to two burets (Figure 3): one filled with Soltrol-130 to the top of the column and another with the aqueous solution to the bottom of the column. The buret with Soltrol-130 was then raised over the course of a couple days until Soltrol-130 no longer entered the column. The change in column weight was then recorded and the water residual saturation, S_{wr}, was determined. Then the column was connected to a syringe pump, and aqueous solution was pushed through the column at 0.3 ml/min until the column weight stabilized. The column weight was again recorded and noted as the point in which the NAPL saturation was at a residual, allowing the organic residual saturation, S_{or}, to be calculated.

Styrene Polymerization Columns. Epoxy and styrene were used in sand columns to obtain a solid thin section of the sand and fluids for inspection under a microscope. Styrene columns were similar to the short columns in design. They had Teflon endcaps and a glass column of similar dimensions, but the caps were bolted to each other rather than individually screwed into the glass. The fluids involved with the experiment were the aqueous solution, a styrene polymer with fluorescent dye, and Tra Bond 2114 epoxy (7,17,19). The porous media was packed into the column and de-aired as described previously in the short column procedure. Once polymerization initiator was added to the styrene, the experiment needed to be completed in less than 18 hours, otherwise the increasing viscosity of initiated styrene would affect experimental results. The water saturated column underwent a NAPL (styrene) flood and a subsequent aqueous flood using the buret and syringe system described in the short column procedure. The styrene was polymerized by placing the column in a pressure vessel, applying a nitrogen atmosphere at 80 psi, and heating the vessel for 40 hours at 85°C. The top of the column was opened and the aqueous solution was volatilized in a laboratory furnace, clearing the pore space of the wetting fluid (7,17,19). Epoxy was then injected into the column to replace the aqueous phase and provide support to the soil column. Once the epoxy hardened, the column was cut open, sliced, polished, viewed under epifluorescent microscope, and photographed. The residual NAPL in the soil could then be observed.

Micromodels.   Glass-plate micromodels were used in flow visualization experiments. The models were made from 10 x 15 cm glass plates with pore networks etched in them (7,17,19). Pore throats and pore bodies were approximately 0.3 and 0.8 mm in diameter, respectively. The model was initially saturated with aqueous solution. A colloidal solution of the Ca-montmorillonite clay in aqueous solution was injected into the model (miscible displacement), and the particles were allowed to settle in the pores with the model oriented vertically. Soltrol-130, dyed red, was flooded into the model followed by an aqueous flood in the opposite direction using the buret and syringe method described in the short column procedure. The liquid phases and the clay could then be observed under a microscope and photographed or filmed as they migrated through the pore network.

Results and Discussion.

The experimental data collected consisted of residual water and NAPL saturations from short columns, photomicrographs of the styrene column thin sections, and photographs of the micromodel experiments.

Short Columns. The short column experiments gave quantitative measures of NAPL saturations as a function of clay content. One hundred and sixteen experiments were done. The residual saturation for NAPL in the uncleaned Sevilleta sand was 22.5±1.6%, higher than the 17-18% expected for a uniform sand, but less than the 27.1± 1.7% found by Wilson et al (7). Wilson's sand and the sand used in this study were collected at different times and on different parts of the dunes, therefore the clay contents were likely different (19). When the fine fraction was cleaned from the Sevilleta sand, the residual NAPL saturation decreased to 16.9±1.1%. The small amount of material removed from the sand had a large effect on the residual trapping of the NAPL. To quantify this effect, residual saturations were determined for different clay contents in the Sevilleta sand. The results are summarized in Table II.

Table II.   Summary of Short Column Results

sample	clay content	porosity	bulk density	S_{wr}[1]	S_{or}[2]
Sevilleta	0.4%[3]	32.5±0.5	1.716±0.014	16.4±1.6	22.5±1.6
cl-sev3-0.0x	0.0%	37.5±0.7	1.656±0.018	8.7±2.1	16.9±1.1
cl-sev3-0.5x	0.5%	36.7±1.2	1.678±0.032	12.1±2.4	20.7±1.4
cl-sev3-1.0x	1.0%	35.6±0.5	1.707±0.014	12.8±1.5	23.8±1.7
cl-sev3-2.0x	2.0%	34.8±0.5	1.728±0.013	23.0±2.7	27.1±1.6
cl-sev3-3.0x	3.0%	34.4±0.8	1.739±0.020	25.6±1.5	30.3±1.6

[1]   residual water saturation
[2]   residual organic or Soltrol saturation
[3]   2.1% of sample was silt and clay

Figure 4 shows the residual water and NAPL saturations plotted against clay content. With increasing clay content, the residual water and NAPL saturations increased. Note that the natural, uncleaned Sevilleta sand, with a clay content of 0.4%, had considerably higher water and NAPL residual saturations than the sandpacks of similar clay content. We believe that the silt sized fraction, 0.002 mm to .04 mm, also acted in increasing the residual saturation of the Sevilleta sand.

Styrene Polymerization Columns. The thin sections from styrene polymerization columns showed a greater NAPL residual for the natural Sevilleta sand than the for clean Sevilleta sand (Figures 5 and 6, respectively). Image analysis revealed the clean Sevilleta sand had a residual saturation of 20.9% and the uncleaned Sevilleta sand had a residual saturation of 32.2%. The difference in these numbers from the short column results was attributed to two-dimensional analysis of three-dimensional data and shortcomings in the styrene column procedure, part of which involves fluids with much

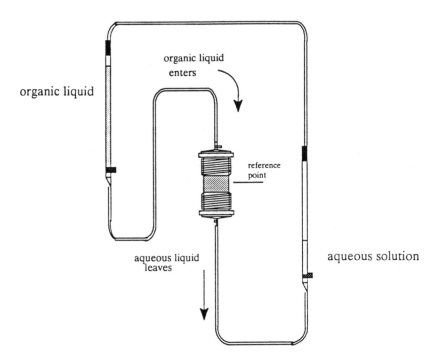

Figure 3. Organic liquid flood on a short column experiment.

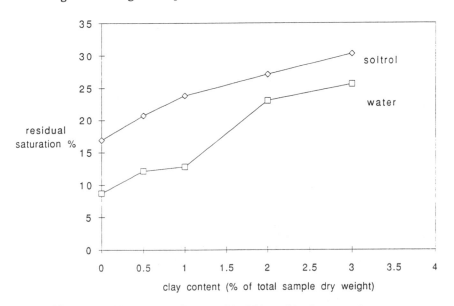

Figure 4. Water and Soltrol (NAPL) residual saturation as a function of clay content.

Figure 5. Microphotograph of the NAPL residual (bright spots) in the indigenous Sevilleta sand. Orientation: vertical cross-section. Magnification: 15×.

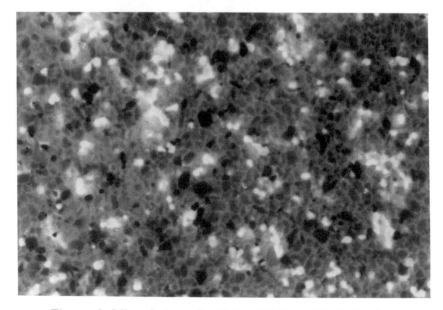

Figure 6. Microphotograph of the NAPL residual (bright spots) in the clean Sevilleta sand. Orientation: vertical cross-section. Magnification: 15×.

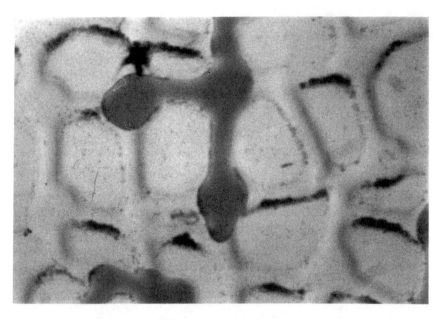

Figure 7. Photograph of a micromodel with NAPL residual.
Magnification: 19×.

greater viscosities and different interfacial tensions (*7,19*). Wilson et al. (*7*) used acids to dissolve the minerals in the sand packs to observe the styrene residual. They noted that the uncleaned Sevilleta sand had more complex ganglia than the clean sands. This indicated that the Sevilleta sand had a greater propensity for by-passing. Yet paradoxically, the image of the uncleaned sand (Figure 5) appears to show many trapped singlets, which is indicative of snap-off. However, observations of uncut columns showed the residual to be complex ganglia. The simple forms found in Figure 5 are believed to be arms of these different ganglia. Unfortunately, clay could not be observed under the microscope because drying of the styrene column to evaporate the aqueous solution collapsed the clay features.

Micromodels. The micromodel experiments were useful in showing how the clay and the fluids interacted in the pores. It is important to note that in these experiments the clay has been dispersed into the pores and then flocculated into small aggregates. These aggregates, comprised of clay particles electrically attracted to each other, can easily block pore throats. If the porous media had a clay content that remained on the sand surface, it is unlikely the same effect on NAPL trapping would be observed.

During a Soltrol-130 flood, the clay was physically mobilized by the liquid-liquid interface. The interface moved through pores and displaced the clay from the pore walls. The clay moved within the aqueous liquid and never crossed the interface into the non-wetting liquid. Wan and Wilson observed a fluid-fluid interface mobilize colloidal particles off pore walls and described the forces involved (*21*). The clay particles could be compacted by the liquid-liquid interface into pore throats, essentially closing the pores from any further fluid flow. Clay particles were trapped between the opposing interfaces of pendular rings. Other clay particles would be compacted onto the sides of the pore walls, commonly in pore throats, changing the capillary properties of the porous media by decreasing the pore throat radius and thus increasing the pore aspect ratio (Figure 7). When the micromodel was flooded with water, large pockets of Soltrol were by-passed leaving behind a considerable NAPL residual. Apparently, clay particles caused greater by-passing by blocking pore throats and increasing the pore aspect ratio.

Conclusion.

A small amount of clay flocculated in the pore space of unconsolidated porous media can have a large effect on the trapping of non-aqueous phase liquids (NAPLs). The Sevilleta sand used in this research had a residual NAPL saturation of 22.5±1.6% but, when cleaned of clay and fine particles, had a residual of only 16.9±1.1%. Experiments in which clay was artificially added to clean Sevilleta sand confirmed this result and showed that residual water and NAPL saturations increased with increasing clay content. Styrene columns dramatically illustrated this difference in residual trapping and suggested that the primary capillary trapping mechanism was by-passing. Micromodel experiments suggested that clay particles in the pores were actively moved by the aqueous phase-NAPL interface. The mobile clays blocked and constricted pore throats, leading to greater by-passing and thus a larger NAPL entrapment. An understanding of how clays affect the trapping of NAPLs can improve remediation efforts, and perhaps prevent a scenario in which a considerable amount of additional NAPL is trapped due to clay content in the porous media.

Acknowledgments.

This research was funded by the Department of Energy Sub-Surface Science Program, contract # DE-FG04-89ER 60829. We would like to thank the reviewers for the excellent suggestions and recommendations and the many people who helped with concepts, experiments, and implementation of this research which include: Jiamin Wan, Mike Wei, Robert Bowman, Norman Morrow, John Hawley, Jacque Renault, William Weiss, Wendy Weiss, Wendy Soll, Anil Bagri, Lisa Carrol, Paul Hoffman, Edith Montoya, Loretta Murillo, and Alan Dutton.

Literature Cited.

1. Frenkel, H, J. O. Goertzen, and J. D. Rhoades. *Soil Science Society of America Journal*. **1978**, *vol 42*, pp. 32-9.
2. Shainburg, I., J.D. Rhoades, D.L. Suarez, and R.J. Prather. *Soil Science Society of America Journal*. **1981**, *vol 45,* pp. 287-91.
3. Alperovitch, N., I. Shainburg, R. Keren, and M.J. Singer. *Clays and Clay Minerals*. **1985**, *vol 33*, pp. 443-50.
4. Gal, M, L.D. Whittig, and B.A. Faber. *Clay and Clay Minerals*. **1990**, *vol 38*, pp. 144-50.
5. Somerton, W.H. and C.J. Radke. *SPE*, **1980**, Tulsa, OK.
6. Yadav, G.D., Dullien, F.A.L., Chatzis, I., and Macdonald, I.F. *SPE Reservoir Engineering*, **1987**, May, pp. 137-47.
7. Wilson, J.L., S.H. Conrad, W.R. Mason, W. Peplinski, and E. Hagan. **1990**. U.S. EPA Report EPA/600/6-90/004, R.S. Kerr Laboratory, Ada, OK.
8. Morrow, N.R. and B. Songkran. in *Surface Phenomenon in Enhanced Oil Recovery*. Plenum Publishing Corp: **1981**.
9. Morrow, N.M., and I. Chatzis. Report DOE/BC/10310-20, Department of Energy, **1982**.
10. Chatzis, I., N.R. Morrow, and H.T. Lim. *SPE Journal*, **1983**, vol 23, no. 2, pp. 311-25.
11. Morrow, N.R., I. Chatzis, and J.J. Taber. *SPE Reservoir Engineering*, **1988**, vol 3, no 3, pp. 927-934.
12. Wei, M. *Hydrology M.S. Thesis H91-3*, New Mexico Institute of Mining and Technology, **1991**.
13. Mohanty, K.K., H.T. Davis, and L.E. Scriven. paper SPE 9406, presented at *SPE Annual Technical Conference and Exhibition*, Dallas, Texas, **1980**.
14. Chatzis, I.N. and F.A.L. Dullien. *Journal of Colloid and Interface Science*. **1983**, *vol 91*, pp. 199-222.
15. Wilson, J.L. and S.H. Conrad. in the proceedings of *Petroleum Hydrocarbons and Organic Chemicals in Ground Water*. *NWWA*: Houston, TX, **1984**, pp. 274-98.
16. Moore, T.F. and R.L. Slobod. *Producers Monthly*. **1956**, *vol 20*, pp. 20-30.
17. Conrad, S.H., J.L. Wilson, W.R. Mason, and W.J. Peplinski. *Water Resources Research,* **1991**, in press.
18. Morrow, N.R. *Chemical Engineering Science*, **1970**, vol 25, pp 1799-1815.
19. Mace, R.E. *Hydrology Independent study* H 90-6, New Mexico Institute of Mining and Technology, Socorro, NM 87801, **1990**.
20. Source Clays Repository. pamphlet, Department of Geology, University of Missouri, 1989.
21. Wan, Jiamin and John L. Wilson. in *Colloid and Interfacial Aspects of Groundwater and Soil Cleanup*. American Chemical Society Symposium Series: **1992**.

RECEIVED January 21, 1992

Chapter 18

Exposure Assessment Modeling for Hydrocarbon Spills into the Subsurface

Sensitivity to Soil Properties

James W. Weaver[1], Bob K. Lien[1], and Randall J. Charbeneau[2]

[1]Robert S. Kerr Environmental Research Laboratory, U.S. Environmental Protection Agency, Ada, OK 74820

[2]Center for Research in Water Resources, The University of Texas at Austin, Austin, TX 78712

Hydrocarbons which enter the subsurface through spills or leaks may create serious, long-lived ground-water contamination problems. Conventional finite difference and finite element models of multiphase, multicomponent flow often have extreme requirements for both computer time and site data when applied to field scale problems. Often, data limitations result in situations where application of complex models is not scientifically justifiable. Simplified models of the separate phase flow of the hydrocarbon and its dissolution into ground water may be appropriate for gaining insight into the significant phenomena, emergency response, or generic simulation for regulatory development. This paper outlines the components of a set of screening models for this problem and focuses on parameter sensitivity. Tabulated values of soil properties are used to model releases in typical soil materials. The availability of standard deviations of parameter values allows assessment of model response with regard to typical parameter variability. This knowledge has important consequences for emergency response applications which tend to rely on tabulated data instead of site specific data. Ultimate interest usually lies with dissolved aqueous concentrations, so the parameter sensitivity is assessed through concentrations predicted for down-gradient wells, as well as other significant aspects of the model results.

Subsurface releases or spills of water-immiscible liquids (the so-called NAPLs or nonaqueous phase liquids) present common and difficult environmental problems for Superfund and other EPA programs (1). Previously, a screening model for L-NAPLs (NAPLs less dense than water) has been developed and is named HSSM (Hydrocarbon Spill Simulation Model (2)). The model is based on a suite of assumptions concerning the physical and chemical phenomena which are associated with hydrocarbon releases. The component modules address the significant aspects of the problem, which are unsaturated zone transport from the surface to the water table, spreading in the

0097–6156/92/0491–00217$06.00/0
© 1992 American Chemical Society

capillary fringe and subsequent unsteady mass transfer into the aquifer, and finally transport by advection and dispersion within the aquifer. This model simulates the transport of the L-NAPL by breaking the problem into three parts, which correspond to three named modules in the model. The modules are the KOPT (Kinematic Oily Pollutant Transport), OILENS (Oil Lens Simulation), and TSGPLUME (Transient Source Gaussian Plume) Models. These modules are based on approximations to the multiphase, multicomponent transport equations. Approximate models are used because:

-all modeling involves approximation and idealization,
-field sites normally do not have adequate data to support complex models, and
-standard numerical methods introduce non-trivial errors into the solutions.

Simple models have the advantage that:

-they have low computational requirements,
-they are generally easy to apply, and
-they give insight into pollutant behavior.

Salient features of each of the modules will be briefly described below. A complete discussion is presented in Weaver and Charbeneau and Charbeneau et al. (both submitted to Water Resources Research, 1991).

Model Basis

The problem of a near surface release of L-NAPL is shown in a conceptual fashion in Figure 1. The liquid moves downward under the combined forces of gravity and pressure. The pressure driving force includes a component which is due to capillary pressure. In the vadose zone, transport is assumed to be one-dimensional downward. The rationale for neglecting lateral transport is that by assuming that all of the mobile L-NAPL moves downward, the model is conservative in the sense that a worst case results when the maximum amount of L-NAPL is available to reach the aquifer and cause ground-water contamination.

    Efficiency is achieved in the KOPT model primarily because the gradients of the capillary pressure are neglected. This assumption causes the transport equation to become hyperbolic, so it can be solved by the generalized method of characteristics (3). In the generalized method of characteristics, the original governing partial differential equation(s) are reduced to systems of coupled ordinary differential equations. Two sets of these are generated: those describing smooth waves and those describing discontinuities. The ordinary differential equations can then be solved analytically for a limited number of problems (4) or numerically by well-known techniques. For KOPT, a Runge-Kutta (5) method is used. One major effect of neglecting the capillary pressure gradient on the simulation results is that the leading edge of the L-NAPL moving into the soil is idealized as a sharp front. Some experimental data show soil profiles which have nearly sharp fronts. Reible et al. (6) show infiltration results for an L-NAPL phase where the fronts are nearly sharp in coarse sands. Similar results have been found in flow visualization experiments conducted in coarse sands at RSKERL. For cases where the front truly spreads, Charbeneau (*4*), for example, presents a theoretical proof that the mean displacement speed of the sharp and true fronts is the same. Smith (*7*) presents a numerical result for water flow showing that a numerical solution of Richards equation was tracked by a sharp front solution. Thus, even if the front is spreading, the kinematic model should be able to capture its mean speed. During infiltration, however, the capillary gradient plays a major role in determining the liquid fluxes in the soil. Therefore, an approximate dynamic correction is added to the model to account for the soil's infiltration capacity. The form of this correction is the Green-Ampt Model (*8*).

In KOPT and OILENS, the L-NAPL is treated as a two-component mixture. The first component, the L-NAPL itself, is assumed to be soluble in water and sorbing. Exposure to infiltrating recharge water and contact with the flowing ground water may cause the L-NAPL to be dissolved. The L-NAPL's transport properties (density, viscosity) and the soil properties (capillary pressure, relative permeability), however, are assumed to be unchanging. The second component is a chemical constituent which can partition between the L-NAPL phase, water phase and the soil. This constituent of the L-NAPL is considered the primary contaminant of interest because, of the hundreds of chemicals which compose typical L-NAPLs, only a few are viewed as causing significant health risks, i.e., benzene, toluene, ethyl benzene and the xylenes. A kinematic approach is used in KOPT and OILENS for solute transport, which results in a model that neglects dispersion. Solute transport is assumed to be the result of advection of water and the L-NAPL only. The chemical is assumed to partition between the phases according to equilibrium, linear partitioning relationships.

Once the L-NAPL is in the vicinity of the water table, radial spreading is assumed to occur. The model ignores the slope of the regional ground water table, primarily for efficiency. When the L-NAPL source is active, however, even with mildly sloping water tables, the flux at the water table is driven radially. Only later, when the source becomes inactive, does the effect of the sloping water table become dominant. In OILENS, when the source strength is reduced, the mound height decays. L-NAPL is trapped both above and below the water table at residual saturations which are allowed to differ between the vadose and saturated zones. By invoking the Dupuit assumptions that the fluxes in the lens are horizontal, and a buoyancy relationship, control volume equations can be developed for the lens (Charbeneau, R. J., et al., submitted to Water Resources Research, 1991). This formulation results in a system of two ordinary differential equations in two unknowns. These two equations can be solved numerically by the same ordinary differential equation used for KOPT. The two models are implemented in one computer code for this reason.

Mass transfer into the aquifer is a function of ground-water flow velocity, lens radius, partition coefficient, recharge rate and other factors. Once in the aquifer, transport is simulated by a quasi-three-dimensional model. Depth, the third dimension, is treated as a penetration thickness; while transverse and longitudinal spreading are simulated explicitly. At some location downgradient, receptor wells can be located and concentrations estimated at those locations. The solution is developed by applying Duhamel's theorem (*9*) to an analytic solution of aquifer transport.

A potential use of the combined model is as a screening tool for development of generic regulations or for emergency response. Both of these activities are normally performed with limited actual data. In the case of emergency response, site data is normally not available during initial phases. Later, data collection may yield enough data for application of a more complete model. Initially, the screening model may be used to assess possible impacts on ground water and to guide data collection. The objective of this paper is to assess the ability of the model to function with limited data and assess the impacts of uncertainty in individual model parameters. To that end, literature data were used to make a suite of model runs with ranges of certain parameters to illustrate the effects on the model results.

Model Data Requirements

Table I lists the parameters required for the model. Also included in Table I is a set of parameter values which describe a scenario for the results presented below. In this example, benzene is the contaminant of interest and is assumed to compose 1.1% by mass of the gasoline. The first group of parameters are based on the geometry of the system and configuration of the release. Usually the details of the release(s) are not known--presenting a primary uncertainty for modeling. In the second group of parameters are the hydraulic properties of the porous media. Values of these parameters have been tabulated based on soil type (10). These provide reasonable estimates for parameter values during emergency response or for generic simulation. By providing average values and standard deviations, the impact of using tabulated data on the model results can be assessed. Table II shows data from Brakensiek et al. (10) for three sandy soil types which are used in the following simulations. Note that in this data set, the loamy sand has higher hydraulic conductivity and lower air entry head than does the sand. Presumably, the sand class is dominated by fine sands. The L-NAPL/water partition coefficient for the benzene is estimated from idealized gasoline component mixtures and ideal solution theory (Raoult's law).

Vadose Zone Results

Of the many possible combinations of effects which can be evaluated by the model, only a very few can be presented in the following sections due to space limitations. Special emphasis is placed on the soil hydraulic parameters which are believed to have a great impact on variation in the model results. A previous presentation (2) also presented limited results for the sensitivity of the model but was limited to one media type and used arbitrary values for the L-NAPL - water partition coefficient. The previous paper also focused on illustrating the character of results from each module. Here emphasis is placed on parameter sensitivity.

The effect of variation in hydraulic conductivity on vadose zone model (KOPT) results is evident in Figure 2. Saturated hydraulic conductivity is used in the model with fluid densities and viscosities to estimate a conductivity for the L-NAPL. The standard deviation of hydraulic conductivity for the average sand of Brakensiek et al. (10) is approximately one order of magnitude. This results almost directly in one order of magnitude variation in front position with time. The other two soil types also show similar variability so this result is typical. The model proves to be much less sensitive to variation in the pore size distribution index λ and the air-entry capillary pressure head (Figures 3 and 4). The variation in the air-entry head is greater than an order of magnitude; but its effect on the model results is far less than an order of

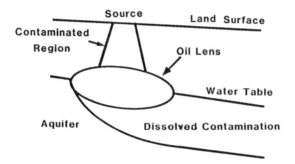

Figure 1 Conceptual representation of L-NAPL release

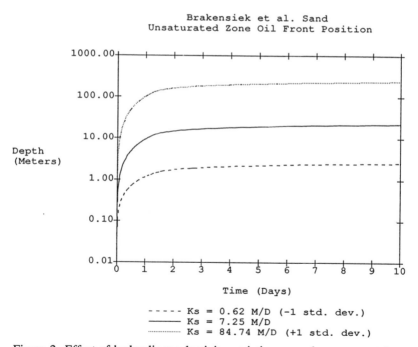

Figure 2 Effect of hydraulic conductivity variation on vadose zone results.

Table I. Required Input Parameters

Geometry and Source Definition

Radius of the source	5.0 m
Depth to the water table	5.0 m
Saturated aquifer thickness	10.0 m
L-NAPL Release:	
L-NAPL ponding depth and duration	0.1 m; (1.0 d)

Hydraulic properties

Saturated hydraulic conductivity, vertical	(7.25 m/d)
Ratio of horizontal to vertical conductivity	5.0
Porosity	0.4
Bulk density	1.6 g/cc
Water density	1.0 g/cc
Water viscosity	1.0 cp
Water surface tension	65 dyne/cm
Maximum water relative permeability during water infiltration	0.5
Brooks and Corey parameters of the capillary pressure curve	
Pore size distribution index	(0.55)
Residual water saturation	(0.049)
Air entry head	(0.173 m)
Groundwater gradient	1%
Aquifer longitudinal dispersivity	10.0 m
Aquifer transverse dispersivity	2.0 m
Aquifer vertical dispersivity	0.1 m
Capillary fringe height	(0.173 m)
Recharge rate or vadose zone water saturation	0.35

Hydrocarbon properties

Density	0.75 g/cc
Viscosity	0.45 cp
L-NAPL surface tension	30.0 dyne/cm
Residual saturation in the vadose zone	0.10
Residual saturation in the aquifer	0.20
Maximum L-NAPL saturation in lens	0.99
Contaminant	
L-NAPL/water partition coefficient	300
Contaminant/soil partition coefficient	0.83 l/kg
Initial contaminant concentration in the L-NAPL	8100 mg/l

Values in parentheses are varied in the simulations that are discussed in the text.

Table II. Selected Soil Data

Soil	Sample Size	$(\lambda)^{1/2}$	ln h_{ce} (cm)
Sand	19	0.739 (0.0170) 0.324, 0.546, 0.826	2.853 (1.174) 5.36, 17.3, 56.1
Loamy Sand	69	0.670 (0.110) 0.314, 0.449, 0.608	2.273 (0.981) 3.64, 9.71, 25.9
Sandy Loam	166	0.615 (0.143) 0.223, 0.378, 0.575	2.820 (1.042) 5.92, 16.8, 47.6

Soil	η	Θ_r	ln K_s (cm/s)
Sand	0.349 (0.107)	0.017	-4.780 (2.458) 7.19e-4, 8.40e-3, 9.81e-2
Loamy Sand	0.410 (0.065)	0.024	-3.818 (1.793) 3.66e-3, 2.20e-2, 1.32e-1
Sandy Loam	0.423 (0.076)	0.048	-5.248 (2.025) 6.94e-4, 5.26e-3, 3.98e-2

Each parameter is given as its mean value in its transformed probability scale; the number in parentheses is the standard deviation in that scale. In the row below the transformed values are the actual values which appear in the order of minus one standard deviation, the mean, plus one standard deviation.

λ = Brooks and Corey's pore size distribution index
h_{ce} = air entry head (Brooks and Corey)
η = porosity
Θ_r = residual water content (Brooks and Corey)
K_s = Saturated hydraulic conductivity

Source: Adapted from reference (*10*).

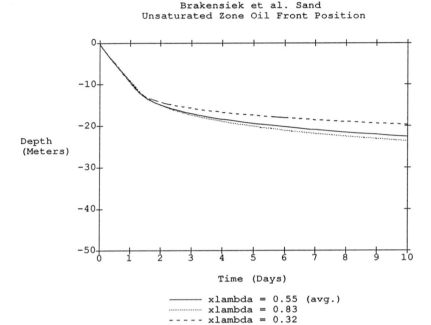

Figure 3   Effect of pore size distribution index variation on vadose zone results.

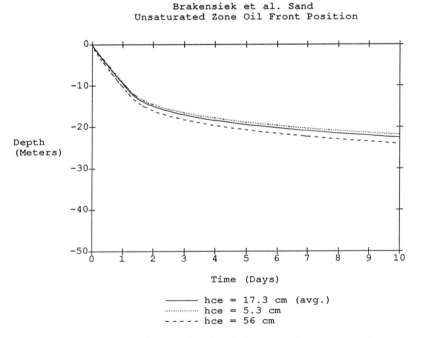

Figure 4   Effect of air-entry head variation on vadose zone results.

magnitude, suggesting that the model is relatively insensitive to air-entry head. These comparisons illustrate that for the average sand, the model is highly sensitive to hydraulic conductivity. Estimates of vadose zone transport using the tabulated data can be expected potentially to be in error by at least an order of magnitude; more if plus or minus two standard deviations from the mean are considered. Other parameters which affect the effective hydraulic conductivity to the L-NAPL are its density and viscosity and, indirectly, temperature. The model is also sensitive to these parameters, through their effect on the effective L-NAPL conductivity.

L-NAPL Lens Results

Considering now the L-NAPL simulations at the water table, results are discussed for the lens radius and the mass flux into the aquifer. As noted in the introduction, the size of the L-NAPL lens at the water table partly determines the boundary condition of the aquifer model. Since the lens size increases initially, so does the size of the aquifer boundary condition. Figure 5 shows results for the average sand, loamy sand and sandy loam soils, under the conditions of the release scenario presented above. Notably for all three soils, the ultimate extent of the L-NAPL is reached in a relatively short time period, which for gasoline under the release scenario is on the order of 20 days. As will be illustrated below, this time period is insignificant compared to the overall duration of the contamination event. Thus, most of the mass transfer into the aquifer (in the context of the conceptual model used here) occurs at a constant lens size.

The variation in lens radius among the soil types is primarily the result of an assumption within the OILENS model that states that radial spreading of the L-NAPL does not occur until a portion of the capillary fringe is filled. By tying the spreading to the capillary fringe height, the ultimate lens radius depends strongly on the capillary parameters of the system. Thus, in the loamy sand, which has the lowest capillary fringe height of the three soils, the L-NAPL must fill up the smallest volume of media before spreading. In the loamy sand, then, is found the largest ultimate lens radius. Lens radii for sand and sandy loam are similar due to their similar capillary fringe heights. The variability of lens size within one soil type (sand) is illustrated by Figure 6. The order of magnitude variation from minus to plus one standard deviation around the mean air entry head produces ultimate lens sizes of 12.4 m, 19.9 m and 31.9 m, again inversely proportional to the assumed capillary fringe height. By comparing Figures 5 and 6, it is noted that the variation within the sand class is on the same order of variation as the variation between the three average soil types (noting, of course, that the materials are similar).

Turning to the mass fluxes into the aquifer, Figure 7 shows the effect of varying the duration of loading from 1 to 4 days in the average sand. In each case, there is a sharp increase in mass flux from the time at which the L-NAPL reaches the water table (which is apparently zero on this figure because of rapid transport through the unsaturated zone) to a peak mass flux, which is followed by a long and gradual reduction in mass flux. The early behavior is due to the relatively rapid initial increase in lens size, while the late behavior is due to gradual leaching of contaminant from the lens. Recall that leaching is caused by ground-water flow under the lens and recharge water infiltrating through the lens. After time periods on the order of 5000 days,

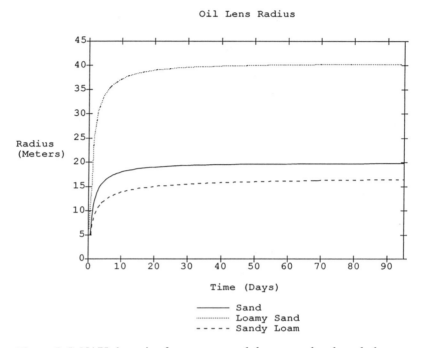

Figure 5 L-NAPL lens size for average sand, loamy sand and sandy loam.

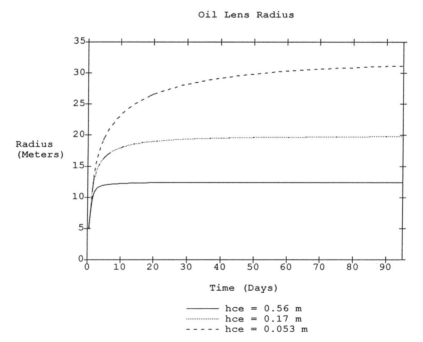

Figure 6 Effect of air-entry head variation on L-NAPL lens size.

almost all of the benzene is leached from the lens. As illustrated in Figure 7, as the duration of the release increases, the peak mass flux increases. The increase is directly related to the increase in lens radius with longer duration of loading, as the larger radii led to larger mass fluxes. There is also a slight shift toward later arrival times of the peak mass flux from 48.8 to 58.2 days, presumably due to the time required for the lens to attain its maximum radius.

When the effect of soil type is included in this analysis, variation in the shapes of the mass flux curves are noted. Figure 8 shows the curves for the three soil types used above with a one-day loading duration. The highest mass flux occurs with the loamy sand since its hydraulic conductivity is the highest. Sand and loamy sand have similar conductivities, and both show lower mass fluxes and correspondingly longer periods of significant flux to the aquifer.

Aquifer Transport Results

Figures 9, 10 and 11 show TSGPLUME results at four downgradient locations for the three average soil types resulting from the mass fluxes illustrated in Figure 8. These results show the pattern displayed in the previous figures, that the loamy sand has the highest hydraulic conductivity and thus the fastest transport times in the aquifer (Figure 9), while the other two types show transport times similar to each other (Figures 10 and 11). The capillary pressure parameters show no direct effect on saturated zone transport, obviously, but may affect the near field behavior of the solution by varying the size of the L-NAPL lens (2). Notably, the concentrations at each well display long tails, an effect which is due to the tailing of the mass flux into the aquifer. The concentrations in the aquifer are generally low as a result of the mass transfer limited flux into the aquifer. Water table fluctuation may be an important, even dominant, mechanism of mass transfer (Johnson, R., Oregon Graduate Center, personal communication, 1991) which is neglected in the current model.

Discussion

The simulations presented in the paper have illustrated the effects of variation in soil properties on spills of hydrocarbon L-NAPLs. From literature data on three soil types, it is apparent that the hydraulic conductivity dominates unsaturated zone and aquifer transport. Due to several assumptions in OILENS, the air-entry pressure is important in determining the size of the lens at the water table and the mass flux into the aquifer. For emergency response or generic simulation, the use of tabulated data introduces large uncertainties into the model results because parameter values may vary by orders of magnitude for a given soil type. These results suggest that the success of such screening will be highly dependent on the quality and availability of the input data. On-going work is focussed on laboratory testing of the vadose zone model for a variety of porous media and L-NAPLs, testing of the key assumptions of spreading in the capillary fringe and mass transfer into the aquifer.

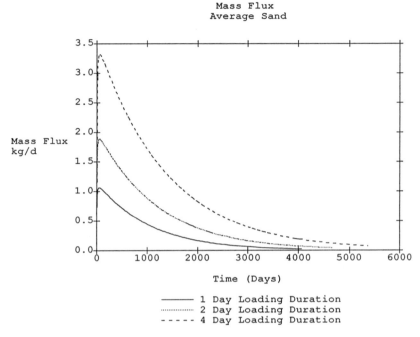

Figure 7   Mass flux variation with variation in loading duration for the average sand.

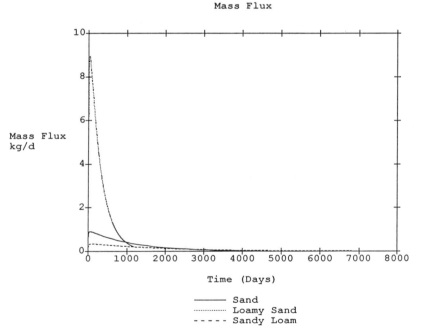

Figure 8   Mass flux variation with variation in soil type.

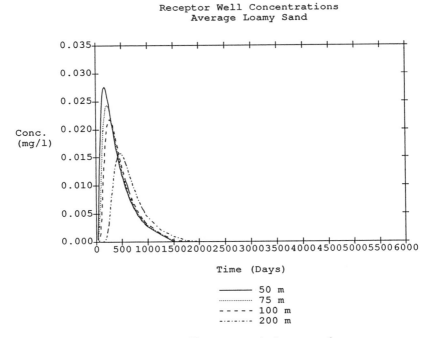

Figure 9 Aquifer transport in loamy sand.

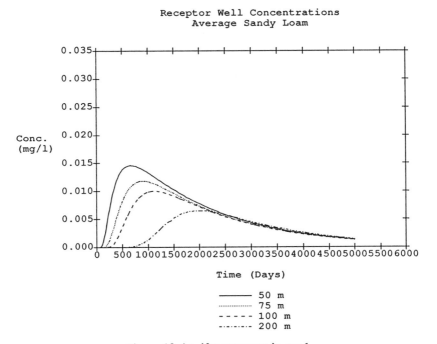

Figure 10 Aquifer transport in sand.

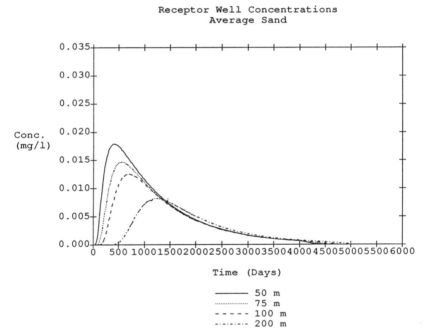

Figure 11 Aquifer transport in sandy loam.

Acknowledgments

Although the research described in this article has been funded wholly or in part by the United States Environmental Protection Agency, it has not been subjected to Agency review and therefore does not necessarily reflect the views of the Agency and no official endorsement should be inferred. All research projects making conclusions or recommendations based on environmentally related measurements and funded by the United States Environmental Protection Agency are required to participate in the Agency Quality Assurance Program. This project did not involve environmentally related measurements and did not involve a Quality Assurance Plan.

Literature Cited

1   *Evaluation of Groundwater Extraction Remedies*, EPA/540/2-89/054, United States Environmental Protection Agency, Office of Emergency and Remedial Response, Washington, D.C., 1989, Vol. 1.

2   Weaver, J.W.; Charbeneau, R.J. in <u>Petroleum Hydrocarbons and Organic Chemicals in Groundwater</u>, National Water Well Association, Dublin, Ohio, 1990, 233-247.

3   Smoller, J. *Shock Waves and Reaction-Diffusion Equations*, Springer, New York, 1983.

4   Charbeneau, R.J. *Wat. Res. Resch.*, 1984, *20*, 699-706.

5   Fehlberg, E. *Low Order Classical Runge-Kutta Formulas*, TR R-315, NASA, 1969.

6   Reible, D.D.; Illangasekare, T.H.; Doshi, D.V.; Malhiet, M.E. *Ground Water*, 1990, *28*, 685-692.

7   Smith, R.E. *Soil Sci. Soc. Am.*, 1983, *47*, 3-8.

8   Green, W.H.; Ampt, G.A. *J. Agr. Sci.*, 1911, *4*, 1-24.

9   Carslaw, H.S., Jaeger, J.C. *Conduction of Heat in Solids*, Oxford, Oxford England, 1959.

10   Brakensiek, D.L.; Engleman, R.L.; Rawls, W.J. *Trans. Am. Soc. Agri. Eng.*, 1981, 335-339.

RECEIVED December 18, 1991

REVIEW AND FUTURE DIRECTIONS

Chapter 19

Transport and Remediation of Subsurface Contaminants
Review and Future Directions

David A. Sabatini and Robert C. Knox

School of Civil Engineering and Environmental Science, University of Oklahoma, Norman, OK 73019

The preceding chapters cover a range of topics within the framework of colloidal, interfacial and surfactant phenomena in subsurface contaminant transport and remediation. The chapters vary from preliminary results of innovative concepts to detailed analyses of important elements within concepts that have been extensively researched. The chapter topics range from modeling to laboratory to field scale studies, and they range from natural transport and fate studies to studies looking at enhanced extraction / treatment concepts, etc. In this final chapter an overview of the preceding chapters will be presented and the session panel discussion will be summarized. In this manner, it will be attempted to summarize where we are (what is the current state of knowledge as presented in the preceding chapters) and where we are going (what were the critical issues delineated and discussed as part of the panel discussion at the symposium).

Where Are We? Overview of Chapters

The preceding chapters have been divided into sections (colloids, inorganics, surfactants, and organics); many chapters could have easily been assigned to several of these categories as the categories are not fundamentally distinct. These divisions have been made in an attempt to aid the reader in assimilating the information in the book. A short discussion of each of these divisions (and the chapters within each division) will be presented below.

The first group of chapters falls into the category of colloids. Colloidal transport in porous media has been extensively studied relative to filtration of drinking water; however, it is only recently that the significance of colloidal transport in the subsurface has been fully realized. Subsurface colloidal transport can greatly increase the migration of compounds associated with the colloids (e.g., strongly hydrophobic organics). This colloidal induced migration can be of great concern if the enhanced migration is not anticipated and accounted for (estimates of migration which ignore colloidal transport will greatly underestimate the

0097–6156/92/0491–0234$06.00/0

contaminant spread and exposure assessment). However, colloidal induced migration can be very beneficial if controlled during remediation (the pump and treat times may be significantly reduced). The chapters in this section discuss fundamental properties of the colloids that may affect their mobility (e.g., nature of colloidal surface, colloidal size, ionic strength), and how to model these processes. The impacts of specific adsorption on the nature of colloids are presented. The adsorption of colloids at gas/liquid interfaces has been described; it was suggested that this may be useful from a remediation approach. The use of polymers to enhance subsurface attachments (and thus immobilize the compounds) has also been presented. Thus, the information in these chapters discusses fundamental properties of colloids, colloidal migration processes, and how to use these fundamentals to enhance remediation (or to render the contaminants relatively immobile and thus reduce risk).

The second division in this book is inorganics. The chapters in the first division dealt with inorganics but in terms of colloids. Much of the initial interest in subsurface contaminant migration and remediation was directed towards organics. It is only recently that inorganics have received increased attention. Treatment of chromate contaminated waters using ultrafiltration and precipitation was discussed. The ability of bacterial strains to reduce contaminants (e.g., selenate) to colloidal forms, which could then be either immobilized in the subsurface or captured in a pump and treat remediation was discussed. In terms of remediation of metals contaminated soils, the use of technologies from the mining industries was discussed, with heap leaching showing the greatest promise. Thus, these chapters were concerned mainly with the remediation of metals contaminated soils and ground waters by either applying old technologies to a new field or by utilizing innovative new processes.

The third division of chapters falls into the category of surfactants. These chapters deal with fundamental properties of surfactants, surfactant-media interactions, surfactant enhanced remediation (in laboratory and field scale studies), and surfactant-biological interactions. The initial chapter in this section discussed fundamental properties of surfactants, surfactant enhanced remediation via solubilization and mobilization, and losses of surfactants in the subsurface (due to processes such as sorption, partitioning into trapped phases, coacervation, precipitation, etc.). Surfactant interactions were discussed in subsequent chapters. The interactions of a range of nonionic surfactants with aquifer materials was presented. The impacts of surfactant sorption on surface and interfacial phenomena was evaluated. The impacts of surfactants on bioactivity was also presented. While it is generally conceived that the surfactants will increase bioavailability of the contaminants and thus increase bioactivity, research presented here suggests that bioactivity was actually inhibited by surfactants (or indirectly inhibited in the presence of surfactants). The use of surfactants to enhance the aqueous concentrations of DNAPLs (solubilization) was discussed; this appears to be a very promising technology for enhancing DNAPL remediation. Results from field scale studies were presented to further substantiate the use of surfactant enhanced solubilization for DNAPL remediation.

The fourth section of this book is entitled organics, dealing with dissolved organics from landfill leachates and nonaqueous phase liquids. The dissolved organics from landfill leachates were observed to impact both the aqueous phase (similar to cosolvation) and the media surfaces; both of these impacts can alter the migration of landfill leachate components. An important parameter relative to NAPLs in the subsurface is the residual saturation of the NAPLs. It was observed that the residual saturation can increase in the presence of fines (colloids, clays). Finally, a simplified model of NAPL migration was presented for use as a screening tool. It is argued that data, hardware and user needs for more complex models are often not justified relative to the quality and availability of input data, thus indicating that a simplified model would be a more useful tool.

In summary, the preceding chapters represent the current state of knowledge relative to colloidal, interfacial and surfactant phenomena in subsurface contaminant transport and remediation. Of course, much additional research is necessary to improve our understanding of these processes. This understanding is vital for optimal management of our invaluable ground water and soil resources. The logical question is where do we go from here. This was the topic of the panel discussion and is discussed below.

What's Ahead? Summary of Panel Discussion

Introduction to Panel Discussion. The culmination of the 1991 ACS Session on Colloidal and Interfacial Aspects of Soil and Ground Water Cleanup was the panel discussion. The focus of the panel discussion was "Where do we go from here?" Having established the state of knowledge (as presented in the preceding chapters and summarized above) and given the significant expertise present, it seemed highly appropriate to consider the critical issues, research needs, etc., relative to ground water and soil remediation. The panel of experts consisted of Dr. Bob Puls, USEPA-RSKERL, Dr. Suresh Rao, University of Florida, Dr. John Wilson, New Mexico Tech, and Dr. John Westall, Oregon State University, with Dr. Westall serving as the panel moderator. The panel discussion was minimally structured, with significant audience participation encouraged. Presented below is a summary of the discussions and significant recommendations of the group (in the order discussed).

Dr. Westall began the discussion by listing areas of research need (grouped into broad remedial categories) which were identified at another conference where several attendees had also been participants. The five areas were as follows: (1) solubilization of hydrophobic organic compounds, (2) mobilization of hydrophobic organic compounds, (3) modification of colloidal stability, (4) immobilization of hydrophobic organic compounds, and (5) alteration of formation permeability. The question put forth to the panel and the audience was "Given limited resources, where do we focus our research efforts (what are the most critical research areas)?" Additionally, the group was asked to indicate the viability of the various remedial categories (relative time necessary to implementation of technologies within each category).

The basic premise behind the remedial categories was alteration of the subsurface physicochemical system.   This indicates the realization that comprehensive remediation of subsurface contamination will require more than physical alterations (excavation, encapsulation, extraction, etc.) to meet regulatory requirements. In order to meet increasingly stringent regulatory cleanup standards in a timely and economical manner, alterations to the subsurface system will be necessary to overcome those physicochemical processes rendering conventional technologies unacceptable.

Although the nation's experts have recognized for some time that remediation of subsurface contamination might ultimately involve introduction of additional agents, the concept has met with extremely limited regulatory acceptance. The reluctance of regulators to allow subsurface injection of remedial agents may be due, in part, to the rather weak promotion of the concept by the research community. However, it is becoming increasingly obvious that realization of significant enhancements in subsurface remediation will require the utilization of remedial agents.

Field Scale Study Sites. A clear mandate from the group discussion was the need for field scale subsurface study sites in the United States (similar to the Borden site in Canada). The viability of field scale study sites in the United States was discussed. Several cases were mentioned where permission was obtained from state regulatory agencies to inject compounds (food grade surfactants, colloids, cosolvents, etc.) in field studies. It was suggested that presenting the need for several field scale study sites to ranking officials (e.g., upper level EPA administrators, members of the Science Advisory Board) may be the most effective means for achieving this goal.

The promotion of field test sites was not made lightly by the panel and audience, but rather indicated careful thought and firmness of conviction. The need was uniformly agreed upon by all of the panel and audience, with comments and suggestions indicating that the concept had been given considerable thought. Many of the comments were the result of observations from and personal experiences with the Borden site. Numerous recommendations were put forth relative to field sites and these included the following:

(1)   Development of multiple, but a limited number of test sites in the US. The multiplicity of sites would allow for varying properties amongst the sites (e.g., hydrogeology, geochemistry). Criticism of having only the Borden site included the fact that it represents only one set of conditions, that it is a very homogeneous (simple) system, and that it is becoming "crowded." The limitation of the number of sites would encourage synergism amongst researchers (mentioned as a strong asset of the Borden site) and maximize use of limited resources.

(2)   Development of field sites that include both contaminated and pristine sites. The contaminated sites will provide a readily available database for testing more advanced concepts. However, boundary conditions are often sketchy at best. For example, the mass of chemical released and the temporal and spatial distribution of the release are typically not known with any confidence at these sites. For this reason, artificial contamination of pristine sites (or portions of a contaminated site that will not be affected

by the contaminant plume for the duration of the experiments) will allow for more controlled analyses of concepts and models.

(3)   Development of field sites for delineation of areas of fundamental research. Experience both in ground water research and other fields (e.g., enhanced oil recovery) has shown that observations made in field scale studies often illuminate critical factors which had not been previously realized in laboratory scale studies. Likewise, fundamental laboratory studies are necessary to give direction to and understand observations from field scale studies. Thus, it is vitally important that laboratory and field studies proceed concurrently.

There was some discussion on potential sites (or sources for potential sites) to be utilized for the field studies. Potential categories of sites that were mentioned included public sites (e.g., Superfund), private sites (e.g., industries), and military sites (e.g., military installations). To summarize, it was indicated that the lack of field scale study sites is a major detriment relative to an improved fundamental understanding of natural subsurface contaminant transport and to the development and implementation of evolving technologies for enhanced subsurface remediation.

Organics / Non-Aqueous Phase Liquids. The topic of discussion next turned to the critical areas of research relative to the remediation of non-aqueous phase liquids (NAPLs). A major obstacle for remediation of NAPL contamination (and especially dense NAPLs--DNAPLs) is the recovery of the residual and free phase fractions (especially the DNAPLs that are in the saturated zone). These fractions are trapped due to capillary and interfacial forces. Remediation using conventional pump and treat methods will require extraction of excessive pore volumes of the ground water to recover the trapped phase(s) due to mass transfer limitations. The use of surfactants can increase this mass transfer by one of two mechanisms: (1) enhanced solubilization increases the apparent aqueous solubility due to the presence of surfactant micelles in the aqueous phase and (2) enhanced mobilization decreases the NAPL-ground water interfacial tension and increases the tendency for emulsions to occur (either macro- or micro-emulsions). Enhanced solubilization will significantly increase the rate of extraction of contaminants from the subsurface. The increased rate of extraction with enhanced mobilization will be even more dramatic. However, some concern exists that enhanced mobilization will result in vertical migration of the DNAPL which may not be captured by the ground water extraction process. In this manner, the problem may actually be worsened. This concern has not eliminated enhanced mobilization as a candidate technology; for example, USEPA is funding research to evaluate the viability of both enhanced solubilization and enhanced mobilization.

Originally the major concern in DNAPL remediation was reducing the time necessary to elute the trapped phases. Current preliminary research suggests that surfactants can effectively address this concern, although fundamental research is necessary to effectively implement this technology. The next level of challenge will be reducing the contaminants to regulatory mandated levels. If the surfactants increase bioactivity in the subsurface system, this could serve as a polishing step in the DNAPL remediation. However, recent research has indicated that, under certain conditions, surfactants may actually hinder bioactivity. Additional research is necessary to understand the interactions between surfactants, contaminants, and microorganisms in the subsurface. Pulsed pumping was also suggested as a way of improving the extraction efficiency. Periods of no pumping would allow the DNAPLs to approach equilibrium concentrations in the ground water prior to restarting the extraction process. It can thus be seen that innovation and imagination are necessary to optimize the remediation of subsurface contamination.

Surfactant sorption on subsurface materials may change the nature of the surface and thus serve to increase sorption to the media (e.g., sorption of surfactant on an anionic surface can produce a hydrophobic surface). This may act to slow the migration of dissolved organic constituents. The impact of this phenomenon on biodegradation of soluble DNAPLs (e.g., chlorocarbons) was considered. One possible scenario is that the retardation of the contaminant, as well as the availability of surfactant monomers as a primary carbon source, may enhance bioactivity. The impacts of the lower aqueous phase concentrations and the surfactants on the biokinetics was questioned.

The above discussions raised the question of how well processes active in the subsurface are understood. As a general rule, it was stated that the physics and chemistry of the processes are better understood in the laboratory than are the biological processes. However, at the field scale it was suggested that the understanding of the physics and chemistry of the system is not significantly better than the biological processes. This again demonstrates the need for field scale study sites. It was also mentioned that the nature of the aqueous chemistry and the nature of the solid media surfaces are becoming increasingly important to understanding subsurface processes. Initial investigations of contaminants in the subsurface focused on neutral organic chemicals in near-surface materials (high organic content). In this system, the hydrophobicity of the chemical and the organic content of the media were observed to dominate the interaction (sorption). The aqueous chemistry (pH, ionic strength) and the nature of the organic content and solid surfaces did not appear to significantly affect the system. The neutral organic compound and high organic content system is fairly simple, and thus promoted thinking of the subsurface as a black box. However, as the system parameters have broadened (compounds of interest are polar, ionogenic organics, metals, complexes, etc.), the nature of the compound, the aqueous chemistry and the nature of the reactant surfaces have realized greater importance.

Inorganics.   The group also discussed the impacts of surfactants on metals contamination in the subsurface. It is possible that surfactant adsorption may displace metals, thus mobilizing the metals. If the surfactant adsorption is intended to serve as a barrier for organic contaminant migration, then the mobilization of the metals may be a negative effect. However, if remediation of the metals contaminated material is an objective, then the mobilization of the metals is a positive effect. Some research has suggested that the adsorption of organic cations (e.g., cationic surfactants) is by ion exchange (as evidenced by little pH effect but greater ionic strength effects) while the adsorption of metals is generally by a surface complexation (pH dependent, ionic strength independent). Surface complexation is a higher energy interaction than the ion exchange interactions; thus, the surfactant adsorption would not be expected to mobilize metals. It was commented that anionic contaminants that undergo the lesser energetic ion exchange reactions may be susceptible to displacement by adsorbing surfactants. The importance of organic and inorganic interactions in the subsurface is amplified by the fact that many sites have both organic and inorganic contamination. The interactions of the mixed contaminants with the media surfaces and each other may amend the behavior of the organics and the inorganics relative to their behavior individually.

The significance of soil (near-surface) versus ground water remediation was briefly discussed. It was suggested that soil contamination may be significantly different from ground water contamination in terms of the magnitude of contamination, feasible remedial responses, etc. For example, feasible technologies for contaminated soils might include excavation with subsequent composting or fines separation (with treatment of the fines rather than the entire matrix). These technologies may not be feasible for soils contaminated in the saturated zone. The impacts of surfactant washing (in situ or above ground) and fines separation on the soil were also discussed (is it still a "soil"?).

Closure of Panel Discussion. The panel discussion was closed by revisiting the five research areas presented at the beginning: (1) solubilization of hydrophobic organic compounds, (2) mobilization of hydrophobic organic compounds, (3) modification of colloidal stability, (4) immobilization of hydrophobic organic compounds, (5) alteration of formation permeability. The group was asked to assess the relative time for implementation of these technical areas. With some discussion, the following assessment in terms of years to implementation were determined: solubilization, 1; mobilization, 2; colloid modification, 5; immobilization, 2; and permeability alteration, 5. It was suggested that these numbers should be viewed relatively (which technologies are closer to implementation, etc.) and not in terms of absolute times to implementation (a multiplier was suggested to convert to actual years).

The authors feel that the panel discussion was an invaluable activity and the discourse and interchange of ideas was excellent. The authors wish to thank the panel members and the audience for their participation. It is hoped that the previous chapters and this overview of the panel discussion will assist in establishing our current state of knowledge and the critical issues to be addressed in the future relative to colloidal, interfacial and surfactant phenomena in subsurface contaminant transport and remediation. Many challenges lay ahead in the effective management (protection and remediation) of our ground water and soil resources. It is hoped that this book will help in some small way to attain this goal.

RECEIVED January 28, 1992

INDEXES

Author Index

Affiliation Index

Subject Index

Production: Paula M. Bérard
Indexing: Deborah H. Steiner
Acquisition: Anne Wilson
Cover design: Sue Schafer

Printed and bound by Maple Press, York, PA